THE HEIDELBERG SCIENCE LIBRARY | Volume I

The Physiological Clock

Circadian Rhythms and Biological Chronometry

Revised Third Edition

Erwin Bünning

The English Universities Press Ltd. London
Springer-Verlag New York Heidelberg Berlin
1973

Professor Dr. Erwin Bünning
Institut Für Biologie,
Universität Tübingen
Tübingen, German Federal Republic

First published in 1958: | "Die Physiologische Uhr"
Second German edition 1963
First English edition 1964
Second English edition 1967
Third English edition 1973

All rights reserved.

No part of this book may be translated or reproduced in any form without written permission from Springer-Verlag.

© by Springer-Verlag OHG., Berlin Göttingen Heidelberg 1958, 1963, and 1964.

© 1967 by Springer-Verlag New York Inc. Library of Congress Catalog Card Number 67-12154.

© 1973 by Springer-Verlag New York Inc. Library of Congress Catalog Card Number 73-77343.

Printed in the United States of America.

ISBN 0-340-18108-7 The English Universities Press Ltd. London
ISBN 0-387-90067-5 Springer-Verlag New York
ISBN 3-540-90067-5 Springer-Verlag Berlin Heidelberg

Preface

As recently as 15 to 20 years ago, to proclaim the existence of an endogenous diurnal rhythm was regarded, even by some well-known biologists, as subscribing to a mystical or metaphysical notion. It is with the Symposium on Biological Clocks (1960) held at Cold Spring Harbor, New York, that an era of intensified experimental work in this field was ushered in.

Since the publication of the second English edition of this book, a few thousand papers on the several aspects of biological rhythms have appeared in various scientific journals. The "University of Sheffield Biomedical Information Project" annually dispatches to its subscribers about 1,200 titles on "Biological Rhythms," out of which more than 600 deal specifically with circadian rhythms.

This rapid and substantial development in knowledge necessitated a complete revision of this book, so that the contents would better reflect current knowledge in the field.

Without the help of Dr. M. K. Chandrashekaran the task of completing the manuscript speedily would have been impossible. Dr. Chandrashekaran devoted much care and attention to a thorough reading of the entire manuscript and made many improvements. I am especially thankful to him for this.

I am also indebted to all those who have collaborated with me in planning and performing experiments and who have stimulated fresh thinking through their discussions. Among them in recent years were Doctors Ilse Moser, M. K. Chandrashekaran, W. Engelmann, and W.-E. Mayer. Last, but not least, I thank Mrs. Brigitte Rätze for her careful and competent handling of the typing and general organization of the manuscripts for both the German and English editions.

Tübingen, December 1972　　　　　　　　Erwin Bünning

The following symbols are used:

LD	light-dark cycle
LD $x:y$	light-dark cycle with x hours of light and y hours of darkness. For example, LD 10:14 means a light-dark cycle with a light period of 10 hours and a dark period of 14 hours.
LL	continuous light
DD	continuous darkness

Contents

Chapter 1. Introduction	1
References to Chapter 1	5
Chapter 2. Endodiurnal Oscillations as the Principle of Many Physiological Time-Measuring Processes	7
a. Examples of circadian oscillations	7
b. Do all circadian rhythms have the same kind of cellular basis?	12
c. Historical development	14
d. Length of the periods	17
e. Heredity	21
f. Loss of overt rhythmicity	24
g. Why do organisms use oscillations for chronometry?	27
References to Chapter 2	28
Chapter 3. Periodicity Fade-Out; Initiation by External Factors	34
a. Necessity of impulse	34
b. Fade-out of the periodicity	34
c. The nature of the initiating stimuli	39
d. Special questions about initiation by light and darkness	41
e. Initiation by low temperature	43
f. Absence of rhythms in early developmental stages	43
References to Chapter 3	46
Chapter 4. Autonomy of Cells and Organs; Controlling Systems	48
a. Independent oscillations in unicellular organisms, tissues, and organs	48
b. Mutual entrainment in plants	51
c. Controlling organs in lower animals	51
d. Controlling organs in vertebrates	55
References to Chapter 4	65
Chapter 5. Temperature Effects	71
a. Temperature and length of period	71
b. Temperature effects during different parts of the cycle	76
c. Setting the clock by temperature cycles	79
d. Influence of low temperature	80
References to Chapter 5	86

Chapter 6. Light Effects ... 89
a. Effects of continuous light ... 89
b. Setting the clock by light-dark cycles ... 92
References to Chapter 6 ... 112

Chapter 7. Attempts toward a Kinetic Analysis: Models ... 116
a. General remarks ... 116
b. Linear and nonlinear oscillations ... 117
c. Effects of reduced energy supply ... 118
d. Harmonic and asymmetric course ... 119
e. Conclusions from phenomena such as fade-out and reinitiation ... 120
f. Effect of synchronizers on a relaxation oscillation ... 123
g. The transients and the response curves ... 123
h. Refractory periods in the circadian cycle ... 129
i. Conclusions ... 132
References to Chapter 7 ... 133

Chapter 8. Attempts toward a Biochemical and Biophysical Analysis ... 135
a. Rhythms in enzyme activity ... 135
b. Earlier reports about chemical effects ... 137
c. Various effective and ineffective chemical factors ... 138
d. The possible role of enzyme rhythms with shorter periods ... 141
e. Biochemical oscillations with longer periods ... 142
f. Role of the nucleus ... 143
g. Role of nucleic acids ... 146
h. The microscopic and ultrastructural approach ... 149
i. Cooperation of several cell constituents ... 149
j. The possible role of membranes ... 151
References to Chapter 8 ... 152

Chapter 9. Adjustment to Diurnal Cycles in the Environment ... 157
a. Synchronization with physical rhythms of the environment ... 157
b. Synchronization with biological rhythms of the environment ... 158
c. Special questions on the time sense of insects ... 159
d. Action directed to a preset time by the "head clock" ... 161
e. The synchronizing factors ... 162
References to Chapter 9 ... 164

Chapter 10. Use of the Clock in Direction Finding ... 167
a. Basic phenomena ... 167
b. Peculiarities of certain species ... 171
References to Chapter 10 ... 174

Contents

Chapter 11. Relations between Circadian, Tidal, and Lunar
Rhythms 177
   a. Endogenous tidal rhythm 177
   b. Lunar rhythms 182
   References to Chapter 11 186

Chapter 12. Control of Diurnal Fluctuations in Responsiveness to
External Factors 189
   a. General remarks 189
   b. Responsiveness to light 189
   c. Responsiveness to temperature 190
   d. Susceptibility to other factors 191
   References to Chapter 12 193

Chapter 13. Use of the Clock for Day-Length Measurement . . 196
   a. Survey of day-length measurements 196
   b. Accuracy and reliability of day-length measurements . . 201
   c. The nature of the time-measuring process 204
   d. The pigment systems 216
   e. Distinguishing between increasing and decreasing day length 218
   f. In retrospect 222
   References to Chapter 13 223

Chapter 14. Pathological Phenomena 229
   a. Disturbances under the influence of nondiurnal rhythms of
      the environment 229
   b. Disturbances by dissociating the rhythms 231
   c. Beats: reinforcement phenomena 234
   d. Damage due to the absence of synchronizing stimuli . . 236
   References to Chapter 14 237

Author Index 241

Subject Index 251

The Physiological Clock

1. Introduction

> That period of twenty-four hours, formed by the regular revolution of our earth, in which all its inhabitants partake, is particularly distinguished in the physical oeconomy of man. . . . It is, as it were, the unity of our natural chronology.
>
> C. W. Hufeland, *The Art of Prolonging Life*. Second English translation, London 1797.

Biologists have long been intensively concerned with the ability of plants and animals to adapt themselves to the spatial conditions of their surroundings. Adaptations to the timing order of their environment have been analyzed less thoroughly, although we know from many ecological observations how astonishing these adaptations can be. They are concerned with subordination to cycles of day and night, to seasonal changes, and even to tides or to alternations of spring and neap tides, which are influenced by the phases of the moon. Such orientations are equally important to both plants and animals.

This book is restricted to the physiological measurement of time, which is brought about (in many cases definitely, in other cases probably) by means of oscillations with periods of *approximately* 24 hours, that is, by the *"endogenous diurnal rhythm."* Such rhythms are now referred to as *"circadian"* (*circa*, about; *dies*, day), following the suggestion of Halberg.

> The term "circadian" is not always used in the same sense. In this book it is restricted to 24-hour rhythms which are able to continue in constant temperature and in the absence of LD cycles. Under such conditions, circadian rhythms no longer *exactly* coincide with 24-hour periods. Circadian rhythms reveal an endogenous component; strictly exogenous diurnal cycles should be called daily rhythms or 24-hour rhythms (Wurtman).

The interest in this subject has increased considerably during recent years, for some of the following reasons.

Progress in methods of investigation. Only recently have laboratory conditions become available permitting a sufficiently accurate measurement of these physiological circadian oscillations or a satisfactory quantitive study of the influence of internal and external factors. Provisions for constant temperature and exact control of light and darkness are particularly important since we know that in some cases experiments can be upset by temperature fluctuations of less than 1°C or by an observation light, even if it is applied for only a few seconds. Such possibilities had been disregarded earlier. Improvements in the available recording devices, in mathematical periodicity analysis and in computer techniques are also important factors in the progress of biological rhythm research.

Recognition of the biological value of the internal clock. We now know that organisms cannot only *indicate* the time of day with the help of their physiological clock, but that they also *make use* of this clock for actual time measurements. For example, some plants and animals do not use the "hourglass principle" when they determine the most suitable time of day for a given process. In other words, sunrise does not initiate a "once only" process resulting after a definite number of hours in a signal which thereby terminates the process. Rather, plants and animals readily "recognize" the benefit of measuring time by means of oscillations (in biological terms: The use of oscillations has distinct selective advantages). The use of oscillations makes it possible to "plan," even for several days. For example, bees can be trained to search for their food at a certain time of day. After they have been offered food for several days at exactly the same time, the bees will continue to look for it at that time even if it is no longer available (Fig. 1). An analogous case in plant life is easy to find. An especially interesting example is ecologically related to the *"time sense"*

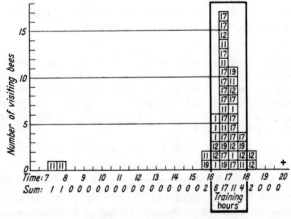

Fig. 1. Time sense in bees. The animals were offered food for several days between 16 and 18 hours (4 and 6 P.M.). Abscissa: time of day and sum of visiting bees. Ordinate: number of the individually marked bees searching for food on the day after the training time. After BELING

of bees. Many flowers open in the morning and offer the insects their nectar or pollen during the day. The secretion of nectar and the production of pollen are both processes with diurnal periodicity (Fig. 105). Opening and closing of flowers is often controlled by the diurnal LD cycles. Obviously, plants possess something equivalent to the "time sense" of bees, because we often see that flowers already begin to open somewhat before dawn, as though they knew that the sun is about to rise. Even if we keep the flowers in the dark, they still open at the same time. In this case we should not speak of a "time sense" although no physiological difference appears at first sight. At all events, plants as well as animals have both correctly measured the course of time, at least during the course of one night.

The study of *photoperiodic responses* has revealed some other very interesting examples of time measurement. It was first discovered in plants, and later in animals, that some developmental processes are controlled by *the length of day*. For example, initiating the formation of flowers, starting winter dormancy of buds, and terminating dormancy in spring can all be regulated by this factor. In animals day length can also control the annual cycle of reproduction or the beginning of rest periods, e.g., the diapause of insects. These photoperiodic reactions are actually caused by *day length* and not by the *quantity* of light. (Of course, a light period of a given length is followed by a dark period of a given length in a normal 24-hour cycle. Which one of the two is more important is irrelevant here.) These photoperiodic reactions can always be produced when the natural day is lengthened by a very weak artificial light. It is not so much the *quantity* and *intensity* of light, as the duration of the light influence that is the important factor in lengthening the light period.

The importance of accurate time measurement by animals was also recognized during the study of another phenomenon, the ability to *orientate by use of a sun compass*. It was found that higher and lower animals can compensate with a high degree of accuracy for changes in the position of the sun, i.e., they can compensate for the progress of time during the day.

Recognition of the medical importance of the internal clock. Intensified interest in the circadian clock was caused by the observation that similar processes of time measurement occur in the human body. These processes are in part responsible for physiological disturbances manifested after a quick transition to an environment in which the cycle phases are shifted. Such disturbances may occur, for example, after flying from east to west or vice versa, or after changing from a day to a night shift. The process of diurnal physiological time measurement is also responsible for diurnal changes in the reaction to damaging external factors, or to drugs, etc.

The sciences of aviation and space medicine show that human test subjects as well as test animals are unable to adjust at will to external cycles (changes of light and dark) that deviate too much from the usual 24-hour cycle. Many other medical aspects could be mentioned here.

Other phenomena. It has long been known that man is still able to measure time physiologically, quite apart from the invention of the clock. Some people are able to estimate with a high degree of accuracy a period of time of several hours even if all external time indicators are missing. These people are able to wake up with only the help of their physiological time measurement just five minutes or less before an alarm goes off (Fig. 2). Even during the daytime this *head clock* allows some people to do everything to a fixed schedule (CLAUSER). This head clock works with greatest precision under hypnosis. The order given by the hypnotist to perform some action after the lapse of a specified time interval will be obeyed with a surprising degree of accuracy.

I have quoted only a few miscellaneous examples. The main purpose of this book is to describe the common features of circadian rhythmicity in plants, animals, and humans. The reader may find literature concerning the more special aspects of botany, zoology, and medicine with the help of the "General Reviews" listed in the "References" to the several chapters.

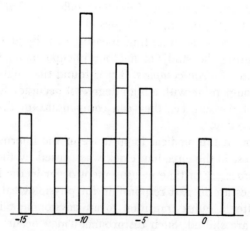

Fig. 2. Intentional awakening of a human individual. Abscissa: 0 = intended time of awakening, negative values mean minutes earlier. Each of the rectangles refers to the awakening at one day. After CLAUSER

References to Chapter 1 [1]
(*Introduction*)

a. General reviews

*Aschoff, J., ed. 1965. *Circadian Clocks*. Amsterdam: North Holland Publishing Company.
*Bierhuizen, J. F. et al., ed. 1972. *Circadian Rhythmicity. Proc. Intern. Symp. on Circadian Rhythmicity*. Wageningen: Centre for Agricult. Publ. a. Documentation.—
*Brown, F. A., J. W. Hastings, and J. D. Palmer. 1970. *The Biological Clock: Two Views*. New York–London: Academic Press.
*Chovnick, A., ed. 1960. *Biological Clocks*. Cold Spring Harb. Symp. quant. Biol. **25**.—
*Cloudsley-Thompson, J. L. 1961. *Rhythmic Activity in Animal Physiology and Behaviour*. New York–London: Academic Press.—
*Conroy, R. T. W. L., and J. N. Mills. 1970. *Human Circadian Rhythms*. London: J. a. A. Churchill.—
*Hague, E. B. ed. 1964. Photo-Neuro-Endocrine Effects in Circadian Systems, with Particular Reference to the Eye. *N. Y. Acad. Sci.* **117**: 1–645.—
*Halberg, F. 1963. Circadian (about twenty-four-hour) rhythms in experimental medicine. *Proc. Roy. Soc. Med.* **56**:253–260.—
*Halberg, F., M. Diffley, M. Stein, H. Panofsky, and G. Adkings. 1964. Computer techniques in the study of biological rhythms. *Ann. N. Y. Acad. Sci.* **115**:695–720.—
*Halberg, F., et A. Reinberg. 1968. *Rythmes circadiens et rythmes de bases fréquences en physiologie humaine*. Paris: Masson et Cie.
*Kleitman, N. 1963. *Sleep and Wakefulness*. Rev. ed. Chicago: Univ. Press.
*Luce, G. G. 1970. *Biological Rhythms in Psychiatry and Medicine*. Chevy Chase, Md.: Nat. Inst. Mental Health.
*Menaker, M., ed. 1971. *Biochronometry*. Proc. of a Symposium. Washington, D.C.: Nat. Acad. Sci.
*Reinberg, A., and J. Ghata. 1964. *Biological Rhythms*. New York: Walker.—
*Richter, C. P. 1965. *Biological Clocks in Medicine and Psychiatry*. Springfield, Ill.: Charles C Thomas.—

[1] General remark on the references: Review articles are also indicated in the text by an asterisk. The references also include reviews which are not mentioned in the text.—In case of French, German, or Italian articles by more than one author, the names are not connected by "and," but by "et," "u.", or "e," respectively.

°ROHLES, F. H., ed. 1968. *Circadian Rhythms in Nonhuman Primates.* Bibliotheca Primatologia No. 9. New York: S. Karger.—
°SEL'KOV, E. E., ed. 1971. *Oscillatory Processes in Biological and Chemical Systems.* Transactions Sec. All-Union Symp. on Oscillatory Processes. 2 Vol. (in Russian with English summaries). Puschino on Oka.—
°SOLLBERGER, A. 1965. *Biological Rhythm Research.* New York: Elsevier.—
°SWEENEY, B. M. 1969. *Rhythmic Phenomena in Plants.* New York: Academic Press.
°WOLF, W., ed. 1962. Rhythmic Functions in the Living System. *Ann. N. Y. Acad. Sci.* 98:753–1326.

b. Other references

BELING, I. 1929. *Z. vergl. Physiol.* 9:259–338.
CLAUSER, G. 1954. *Die Kopfuhr.* Stuttgart: F. Enke.
WURTMAN, R. J. 1966. *Science* 156:104.

2. Endodiurnal Oscillations as the Principle of Many Physiological Time-Measuring Processes

Il n'est point nécessaire pour ce phénomène qu'elle soit au Soleil ou au grand air, il est seulement un peu moins marqué lorsqu'on la tient toujours enfermée dans un lieu obscur, elle s'épanouit encore très sensiblement pendant le jour, et se replie ou se resserre régulièrement le soir pour toute la nuit. . . . La Sensitive sent donc le Soleil sans le voir en aucune manière.

M. DE MAIRAN (on diurnal leaf movements). Acad. Roy. Sci. Paris 1729, p. 35.

(Diurnal movements continue in DD. The sensitive plant senses the sun without seeing it in any way).

a. Examples of circadian oscillations

The examples of physiological time-measurement mentioned in Chapter 1 might still be explained by the principle of an hourglass: one may imagine that a particular event (e.g., sunrise) initiates a physiological process which requires a certain length of time, determined by inheritance.

Demonstration of a cyclical nature. Bees can be trained to search for food at a certain time of day. If they have to remain in the hive for several days (for example, because of bad weather) they still "know" at what time they have to look. They will return to the feeding place at the usual training time (WAHL). Plants can react the same way: flowers (of certain species) kept in darkness for several days still close their petals in the evening and open them in the morning (Fig. 3). In spite of DD they "know" for several days at what time the sunrise is to be expected. This is evidence that the physiological clock works on the basis of an endogenous rhythmicity.

Further examples. The endogenous circadian rhythm is also a factor of many other physiological activities, not only of the "time-sense" of bees or flowers. The action of the physiological clock can be compared with a master clock controlling many subordinate clocks. In other words, it influences many peripheral physiological activities, causing them also to *indicate* the time, being likewise diurnal by

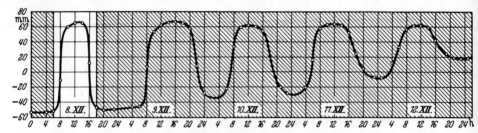

Fig. 3. Continuation of petal movements in *Kalanchoe blossfeldiana* in DD (shaded area). There is only slight damping. Ordinate represents petal movement with rising values indicating flower opening and falling values flower closing. (The absolute values refer to the recording system.) After BÜNSOW

nature. Thus, in order to study this clock, we can choose from many different reactions which are all under control of the same clock.

> The term "master clock" does not mean that only certain cells, tissues, or organs are oscillating autonomously. It only means that we have to distinguish between the overt circadian oscillations and their (hitherto unknown) biochemical or biophysical basis.

Besides the movements of petals, the diurnal up- and downward *movements of leaves* have also been studied (Figs. 4 and 5). These movements are based either on antagonistic differences in the rate of growth of the upper and lower leaf side or on antagonistic fluctuations in turgor pressure in the upper and lower half of leaf joints (*BÜNNING 1958).

In certain fungi and algae, processes such as the *discharge of spores* are easily measured diurnal phenomena which also continue endogenously under constant conditions (SCHMIDLE; UEBELMESSER; INGOLD and COX; BÜHNEMANN; WALKEY and HARVEY; AUSTIN).

In animals we can measure the diurnal *migration of pigments* (Fig. 6) or variations in *running activity* (Fig. 7). The *emergence of insects*

Fig. 4. Bean seedling (*Phaseolus coccineus*). Position of primary leaves at night (left) and during daytime (right). In kymograph records, both with *Canavalia* and *Phaseolus* (Fig. 5 and many others), the highest points indicate night position, i.e., maximum lowering of the leaves.

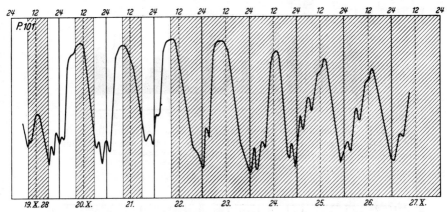

Fig. 5. *Canavalia ensiformis.* Leaf movements in LD cycles and in DD. Dark periods shaded. On October 18, an inverse LD cycle was initiated (light at night and dark during the day). Leaf movements are shifted 12 hours under these conditions and phase-shifted cycle continues in DD. After KLEINHOONTE 1929

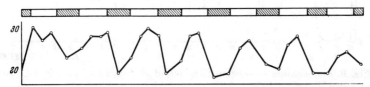

Fig. 6. *Ligia baudiniana.* Periodic variation in pigment dispersion, shown by groups of isopods, kept in the laboratory in DD. Maximum average pigment dispersion would be plotted as 46, maximum concentration as 0. The alternating areas of gray and white correspond to the natural 24-hour succession of night and day. After KLEITMAN 1940

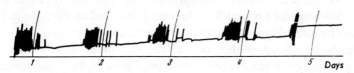

Fig. 7. *Mesocricetus auratus* (hamster). Rhythm of running activity in LL, dim light. Original

from their pupae, which is related to the time of day, is also quite often used to study circadian rhythm (Fig. 8; °REMMERT).

Quantitative fluctuations in *metabolic rate* can be observed in plants and animals under constant conditions (Fig. 9; see also WILKINS; WARREN and WILKINS). Equally striking are diurnal fluctuations in the *photosynthetic capacity* of plants, when measured under constant con-

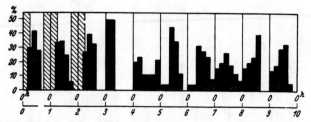

Fig. 8. *Drosophila.* Emergence of flies from pupae. Abscissa: days (0 = midnight). Ordinate: percent of daily total of imagoes that emerge in any given 4-hour period. Dark periods shaded. After BÜNNING 1935a

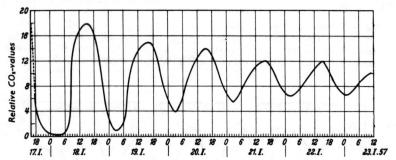

Fig. 9. *Bryophyllum calycinum.* CO_2 output of detached leaves in DD. After WOLF from °BÜNNING 1958

ditions (higher plants: CLAUSS u. SCHWEMMLE; JONES and MANSFIELD; unicellular algae: °HASTINGS and SWEENEY; PALMER *et al.*; °SWEENEY). Fluctuations in the dark fixation of CO_2 (WARREN and WILKINS; WILKINS 1967) as well as in O_2-consumption (MIYATA; BENNETT and GUILFORD; TWEEDY and STEPHEN) are known. For circadian rhythms in CO_2-release by animals see LEVENGOOD, RENSING, and USHATINSKAYA.

An especially interesting and striking rhythm is directly connected to a periodicity of metabolism—the diurnal *periodicity in luminescence* of the unicellular alga *Gonyaulax*, which continues in DD (Fig. 10, °HASTINGS *et al.*; °SWEENEY 1965, 1969). Cellular extracts taken from *Gonyaulax* cells at different times during the day were found to contain varying concentrations of luciferin and luciferase (the enzyme responsible for luminescence). The concentration of these substances was higher during the night than during the day.

Fig. 11 shows examples of diurnal processes in the human body which also continue under constant conditions.

There is evidence that "for the greater part of the 24-hour period

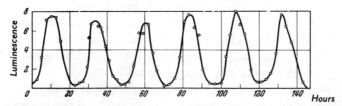

Fig. 10. *Gonyaulax polyedra.* Rhythm of luminescence from cultures maintained in LL (dim light) and constant temperature. After *Hastings and Sweeney

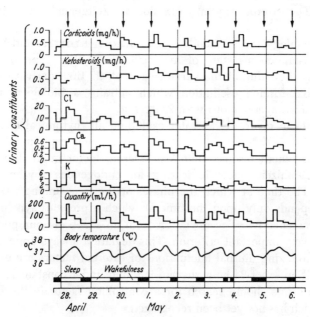

Fig. 11. Periodic processes in the body of one human subject kept without timepiece in complete isolation from the outside world. Cl^-, Ca^{++} and K^+ in milliequivalents/hour. After *Aschoff and Wever

many, if not all of the physiological processes of the organism are in a state of continuous change" (Colquhoun).

Coupling and uncoupling. Processes such as those mentioned above (and many others) can be controlled by the circadian clock (i.e., coupled to the clock). Yet in the same species, but under different conditions, they may proceed independently of this kind of control. For example, in some species the opening and closing movements of the flowers are controlled entirely by external factors (by LD cycles or by alternating high and low temperatures). In some species, mitotic cycles may be coupled to the clock and thus display all the characteristics of circadian rhythms. In other species or under other conditions they are quite independent of it. The mating activity in a stock of *Paramecium aurelia* can be coupled to the clock, but uncoupling by external factors is quite easy (KARAKASHIAN).

Phase position. Although it is quite obvious, it should perhaps be mentioned again that the phases of the endodiurnal rhythm are determined by external factors. The LD cycles or cycles of high and low temperature have a synchronizing effect on the circadian rhythm: they set the clock and adjust the circadian cycles to the exact 24-hour periodicity. The external rhythms function as "Zeitgeber" (cues, synchronizers).

Terminology (recommended by *ASCHOFF, KLOTTER, and WEVER), *Phase:* instantaneous state of an oscillation within a period, represented by the value of the variable and all its time derivatives. *Zeitgeber:* that forcing oscillation which entrains a biological rhythm. *Synchronizer:* synonymous with *Zeitgeber*.

Sources of error. Even after excluding all rhythmical changes of illumination and temperature it still took much experimentation to determine whether some unknown external factor might function as the synchronizer. Facts indicating an endogenous rhythm are as follows. The time at which the phases of the free-running periodicity occur depends only upon the time at which a previous synchronizing factor was effective. The deviation from an exact 24-hour periodicity is also a very important criterion. This deviation always appears when all synchronizing diurnal fluctuations of external factors are excluded. Indeed, if an exact 24-hour rhythm is found under constant laboratory conditions, one should search for a controlling factor in the environment which has not yet been recognized.

b. Do all circadian rhythms have the same kind of cellular basis?

General remarks: "wheels" and "hands." The discovery of endogenous diurnal rhythms led many workers to hypothesize that the overt cycles themselves are feedback systems, or comparable with the wheels of a clock. A few examples may show this.

In earlier papers on endogenous diurnal leaf movements the assumption was made that the downward position of the leaf was the cause for an overshooting upward movement, the overshoot being the direct cause for the ensuing downward movement with an overshoot, etc. PFEFFER (1911) devoted a special paper to this question. He prevented leaf movements by attached weights or some other mechanical resistance. After releasing the leaves and thus allowing them to resume the movements, they did so without a phase shift, when compared with the controls. Thus, PFEFFER came to reject the simple feedback hypothesis and to assume controlling diurnal processes which are going on even when the movements themselves were prevented.

An analogous zoological example is described by *RICHTER. He recorded the running activity of a blinded rat. The cycles measured 23 hours and 45 minutes. Convulsions produced by electroshock made the rat inactive for ten days, after which the animal resumed activity at exactly the predicted time "as if nothing had been done to the rat." Thus, activity is not a "wheel" of the clock.

> Similar experiments were performed by other research workers who studied other metabolic rhythms in plants as well as in animals. For example: High quantities of sugar and high quantities of glycogen in the liver are not the causes for synthesis and hydrolysis, respectively, of glycogen in the liver. The circadian rhythm of glycogen deposit in the liver persists even under conditions of starvation.

More recent research indicates that many research workers preferred or still prefer the other extreme hypothesis: to assume that *all* the overt circadian rhythms are nothing but hands of one and the same master clock in the cell, i.e., to assume an identical cellular or even molecular basis for all these rhythms. This hypothesis seems to be wrong, too.

Developmental cycles. Apparently many developmental rhythms may be explained without assuming a control by the circadian clock. This may be true for several cell cycles which have now been described in detail as part of the intensive studies on cell synchronization (*CAMERON and PADILLA; *ZEUTHEN). Many of these cycles have periods which approach 24 hours, but may vary between about 12 and 30 hours or between more extreme values. In other cases, mitotic cycles may be strongly coupled to the endogenous diurnal periodicity. Several developmental cycles in fungi also belong to those processes which become rhythmical in nature because they themselves have "wheels" of a clock, without being coupled to a master clock which runs independently of these overt morphogenetic events. This may be true, for example, for growth rings in several species of fungi. In *Neurospora crassa* these growth rings are due to an alternation of surface growth and formation of aerial hyphae (BIANCHI). Ring formation in this and other species, or other cycles in fungi such as bioluminescence (BÜHLER u. BÜNNING) continue during constant conditions with periods of about 24 hours. These periods, under certain conditions, may deviate very much from circadian (from less than 20 to more than 100 hours), indicating that we are not dealing here with a control by a cellular mechanism of the type that is responsible for most of the more typical circadian rhythms. Other indications (concerning certain species) are the strong dependence of the period on the chemical composition of the growth medium, a strong influence of temperature on the periods, and the fact that (in *Neurospora* and *Ascobolus*) it is even possible to set the phase by transfer to fresh medium (BERLINER and NEURATH 1965a, b; JEREBZOFF 1961 and *1965; SUSSMAN et al.; BOURRET et al.). Influences such as these are not known for the "classical" examples of circadian rhythms.

But the situation is more complicated. Mitotic cycles and other cell cycles may show certain features in common with the more characteristic cases of circadian rhythms. To these common features belong the synchronizing effects of LD cycles and relative temperature independence, at least under certain conditions (for details see *ZEUTHEN; *BRUCE; *MITCHISON). In other species or under other conditions, such developmental cycles certainly are coupled to the "genuine" circadian clock. This may be true not only for mitotic cycles, but also for sporulation cycles in fungi (SARGENT et al.). In other words, there may be certain phases ("gates") which "permit" the respective event, like mitosis, for instance (EDMUNDS).

Finally, as may be expected, the different methods for biological control of diurnal rhythms may interfere with each other. A very simple example may demonstrate this; the strength of hunger stimuli and time of occurrence depend on the quantity of food we eat. In addition, these stimuli may recur independently or become especially strong at normal meal times even without eating for several days, because of control by endogenous circadian rhythms. An analogous situation exists with respect to sleep and wakefulness, as with many other physiological rhythms.

c. Historical development

Plants. The endogenous diurnal rhythm was discovered first in plants by studying the circadian leaf movements. Such up- and downward movements of leaves are frequent in the plant kingdom, but they are most obvious in the Papilionaceae. During the march of ALEXANDER THE GREAT these movements were observed by ANDROSTHENES on various Papilionaceae (especially *Tamarindus indicus*) (see BRETZL). The experimental demonstration of an endogenous component in diurnal leaf movements was first indicated by the astronomer

DE MAIRAN in 1729. He found that these movements will also continue in DD, and he even points out the possible relation of this phenomenon to certain aspects of human behavior. His intuitive speculations were thus the predecessors of those of HUFELAND (see quotations in Chapter 1 and Chapter 14). Leaf movements were again investigated by ZINN (1759). He confirmed that these movements will continue in the absence of LD or temperature cycles.

More than a hundred years ago SACHS produced good evidence for the existence of hereditary rhythms. SEMON (1905, 1908) emphasized the hereditary nature of the diurnal periodicity. His experiments greatly influenced PFEFFER's work on diurnal leaf movements, begun in 1873. At first PFEFFER rejected the assumption of the existence of hereditary rhythms. His later works, however, contributed considerably to their investigation. In his experiments and those of several others, the diurnal movements of leaves and petals remained the most important tool in the study of endogenous diurnal rhythms. Yet a few authors observed endodiurnal fluctuations in growth rates (BARANETZKY 1879) and other plant processes as early as the last quarter of the previous century and the first years of this century.

Although the experimental data did convince PFEFFER of the existence of an endogenous diurnal rhythm (1915), some later authors were still doubtful. One of the sceptics was ROSE STOPPEL. A counter-argument was the observation that the extreme positions in leaf movements in DD occurred very often at certain times of day. For example, the extreme downward position of leaves was observed frequently a few hours after midnight (Fig. 12). Between the years 1928 and 1932 (BÜNNING u. STERN; KLEINHOONTE) the reason for this was found: the starting time of the experiment influenced the phase positions. In many of the older experiments the decisive effect of a synchronizer ("*Zeitgeber*") had remained unnoticed: usually the experiments were set up with the help of a red safe-light, which at that time was considered physiologically ineffective for plants. But this very light quality is the most important synchronizer for the endodiurnal rhythm of higher plants. We also know that between one or two minutes of light per day (or even shorter light signals) can be very effective cues in plants and animals. KLEINHOONTE, again studying leaf movements, clearly demonstrated the laws of this regulation and also the participation of the endogenous rhythm. She emphasized the random distribution of the phase-positions in DD as an argument for the absence of any external synchronizer. At the same time, the deviations from an exact 24-hour periodicity in DD had been pointed out by BÜNNING u. STERN who measured periods of 25.4 hours in *Phaseolus*.

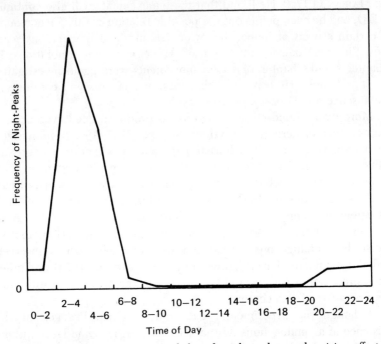

Fig. 12. Example of experiments believed to show the synchronizing effect of "factor X." Diurnal leaf movements of *Phaseolus* in a darkroom. Night peaks (maximum of downward movement) occur mostly about 3 hours after midnight in spite of "constant" conditions.

For further information and additional references see *BÜNNING 1958, 1960.

Animals. During the last century some observations were made on apparent endodiurnal changes in running activity. Further studies on endogenous activity rhythms of mammals were published by C. P. RICHTER (1922); JOHNSON (1926); HEMMINGSEN and KRARUP (1937). In 1894, KIESEL described a fluctuation in pigment migrations in arthropods which continued even when there were no LD cycles (also DEMOLL 1911).

Quite early, several authors investigated the daily course of body temperature in vertebrates and human beings (see ASCHOFF 1955). In 1910, FOREL suspected the existence of a time sense in bees. Proof of the existence of an endodiurnal rhythm was obtained from experiments demonstrating the glycogen rhythm in the liver (FORSGREN 1928, 1935, Fig. 13), the time sense in bees (BELING 1928; KALMUS 1934), pigment migration in Crustaceae (WELSH 1930, see *WELSH 1938), several studies on the diurnal rhythm of emergence of insects from

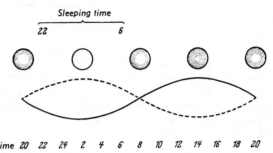

Fig. 13. Periodic changes in liver activity according to FORSGREN. The circles represent liver lobules. The white region indicates the anabolic phase, the dotted region the catabolic phase. Broken line: changes in glycogen content. Solid line: bile production. Abscissa: time of day.

their pupae (BREMER; *KALMUS 1935; BÜNNING 1935a), and on diurnal fluctuations in the activity of Orthoptera (LUTZ 1932). In 1937, BUCK described a diurnal rhythm in the luminescence of the fire fly, *Photinus pyralis,* in continuous dim light.

> Summaries relating especially to older literature are given by *KALMUS (1935), *CASPERS (1951), *WELSH (1938), *PARK (1940), and *JORES (1937).

Humans. Clear indications for the existence of diurnal physiological oscillations in humans became available some time ago through the investigations on the physiology of sleep and wakefulness (*KLEITMAN). Among the diurnal functions in the human body, the rhythm of urinary excretion was investigated quite early, as were temperature and pulse frequency (QUINKE 1893, and several earlier authors mentioned by FUNCK). GERRITZEN pursued these problems further, but only more recent investigations have proved the participation of an endogenous diurnal rhythm.

> The longest known phenomenon of physiological (or "psychological") time measurement is the ability of many people to awake deliberately at a set time or to be always punctual during the day without external cues. CLAUSER summarized the older literature on these phenomena, which had often been considered as something mystical. We lack a clear demonstration that this head clock works on the basis of the endogenous diurnal rhythm.

d. Length of the periods

Degree of variation. In early work, the persistence of a diurnal rhythm under conditions of constant temperature in LL or DD was

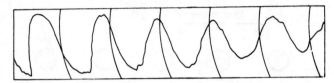

Fig. 14. *Phaseolus coccineus*. Diurnal leaf movements in LL. Guide lines 24 hours apart. Over a period of 6 days there was a phase shift of 17 hours when compared with the normal 24-hour day. Therefore, the period was approximately 27 hours. After BÜNNING u. TAZAWA

repeatedly explained by implicating some unknown external factor. However, the fact that the phases could be shifted to any time of day favored the interpretation that the diurnal rhythm was endogenous. Another, more convincing argument is the fact that in the absence of external synchronizing factors, the period of physiological rhythm shows deviations from the 24-hour period.

Usually plants and animals maintain an average period of 22 to 28 hours. Studying the diurnal leaf movements of beans (*Phaseolus coccineus*), a period of about 27 hours (Fig. 14) was measured, under certain conditions even a length of 28 hours (LEINWEBER). The alga *Oedogonium*, which releases spores in daily periods, shows a cycle of 22 hours (BÜHNEMANN). BALL et al. (1957) measured the rhythm in growth rate of oat coleoptiles (*Avena*) and observed periods of 23.3 hours. The period in animals seems to be in most cases between 23 and 26 hours. It is seldom that the free-running periods do not differ significantly from 24 hours (LOWE et al.).

Constancy, accuracy. From day to day, under constant conditions (LL or DD) an individual plant or animal usually maintains, with relative constancy, a period which is characteristic of this particular individual under a given set of conditions (Fig. 15). The deviations are usually less than one hour or even less than 15 to 20 minutes. If the specific period of an animal or plant is given in fractions of an hour (e.g., 23¾ hours), it still represents a value which is experimentally reproducible. By studying the behavior of an individual for several days under constant conditions, it is often possible to calculate the period with an accuracy of a few minutes (*PITTENDRIGH; *PITTENDRIGH and BRUCE). Especially in rodents, several authors (e.g., DECOURSEY, 1960, 1961) have found values that can easily be reproduced with exact methods. It is possible, then, to state the period length for the particular individual with an accuracy of ±1 to 2 minutes. On the other hand, abrupt or gradual transition to another period may occur because of internal conditions of the individual (ROBERTS; PALMER; LOHMANN; ESKIN).

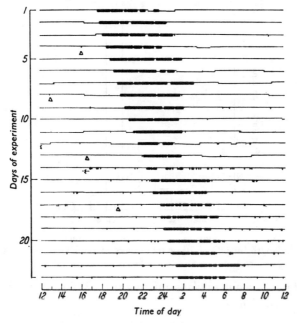

Fig. 15. Activity record of a flying squirrel (*Glaucomys volans*) in DD conditions, showing a rhythm of 24 hours and 21 minutes ±6 minutes. The activity over a period of 23 consecutive days is represented on the 23 horizontal lines. After DeCoursey 1961

The length can also depend upon more overt changes of the physiological condition. This is indicated by observations by Menaker in the bat *Myotis lucifigus:* the cycle of its body temperature measured in an environment of 3 to 10°C shows a period of 22 hours and 25 minutes in the summer, but 25 hours in the winter.

The accuracy with which animals maintain a period is especially remarkable, since the precision of the human head clock had caused some mystical speculations earlier. Thus, Clauser cites the opinion of Bramwell (London 1843): While the normal consciousness is asleep, some sort of intelligence must have observed the course of time.

Individual differences. By comparing several individuals of one species, one can easily find individual differences amounting to one hour, or in some cases even to several hours. This was found in plants as well as in animals and humans. In certain mice, the following specific period lengths were measured: 25.0, 25.1, 25.3, 25.4, and 25.5 hours (Aschoff 1955a). In lizards Hoffmann (1955) found individual differences ranging from 21.1 to 24.7 hours. Fig. 16 gives the variability of

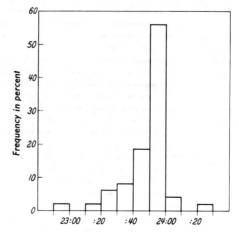

Fig. 16. *Glaucomys volans* (flying squirrel). Frequency distribution of the average cycle length for DD of 10 or more days. After DeCoursey 1961

the period in *Glaucomys volans,* ranging from 23.0 to 24.5 hours, mostly from 23.5 to 23.59 hours.

Aschoff u. Wever (see also *Aschoff 1965) observed humans in bunkers where all external time indicators had been excluded. They found the periods of functions such as those recorded in Fig. 11 to range from 24.7 to 26.0 hours. This is close to values measured by Halberg et al. (Reinberg et al.). Of course these data apply only under certain conditions. In humans too there are external and internal factors by which the length of the period can be influenced.

It should be mentioned that the individual differences occur not

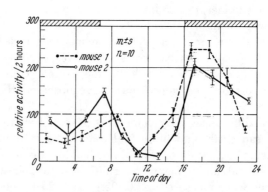

Fig. 17. Individual pattern of activity in two mice, 10 consecutive days superimposed. Abscissa: time of day. Ordinate: relative activity, 100% of ordinate = average activity per 24 hours. Dark periods shaded. After Aschoff u. Honma

only in the length of the periods, but also in the pattern of the curves (Fig. 17).

One should not draw specific conclusions about the *endogenous* diurnal rhythm from curves which were recorded in LD cycles. For example, under diurnal LD cycles two peaks may appear in the running activity (one in the morning, and one in the evening). Under constant conditions this can change to a monophasic rhythm, as was observed with the dung-beetle (*Geotrupes silvaticus*) by GEISLER. In other cases, the two peaks may persist in constant conditions (ASCHOFF 1966, with respect to birds).

Also individuals within one species may vary in this respect, as was observed by NOWOSIELSKI and PATTON in *Gryllus domesticus*. Some individuals showed one daily peak of locomotor activity, others showed a second peak.

e. Heredity

The periodicity . . . is to a certain extent inherited.

CH. and F. DARWIN, The Power of Movement in Plants. London 1880, pp. 407–408.

Modifications? The question has often been raised as to whether the endogenous diurnal rhythm really is inherited or whether it is imprinted by diurnal fluctuations of the environment during an early stage of development. To test this, plants and animals were kept under constant conditions of temperature and light from their earliest stage of development onward, sometimes for several generations. In other experiments, an attempt was made to see if a different rhythm, e.g., 8:8 hours, could be forced upon plants or animals during the early stages of their development by exposing the subjects to corresponding LD cycles. But all these experiments failed, the plants and animals retained their own individual rhythm with cycles ranging from 22 to 28 hours in complete independence of the previous treatment. There was no change even if the preceding generation was treated.

Some examples may illustrate this: seedlings were pretreated with LD 8:8 hours. By recording the diurnal movements of their leaves it was established that there was no modification in their rhythm (KLEINHOONTE 1932). Exposing the mother plants to abnormal LD cycles or to constant conditions has no effect either; the following generation still shows diurnal leaf movements (BÜNNING 1932). Bees maintain their normal sense of time even after being brought up under constant conditions (WAHL 1932). *Drosophila* flies kept under constant conditions from their larval stage on emerge with regular diurnal rhythm (BÜNNING 1935a). The same was found in nightmoths (PARK and

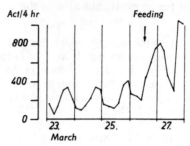

Fig. 18. Activity of freshly hatched chickens. After ASCHOFF u. MEYER-LOHMANN

KELLER 1932; PARK 1935; HORSTMANN 1935). Eggs of chickens and lizards kept under constant conditions produce animals which later on show regular diurnal cycles (ASCHOFF u. MEYER-LOHMANN, Fig. 18; HOFFMANN 1957. Rats were exposed to abnormal LD cycles (8:8 hours), without the rhythm of the following generation being disturbed (HAMMINGSEN and KRARUP). Nor did abnormal LD- or temperature-cycles (9:9 or 18:18 hours) during hatching of chickens or lizards modify the subsequent rhythm of activity of the animals (HOFFMANN 1959). *Drosophila* cultivated for 15 generations under dim continuous light did not lose the capacity to emerge from their pupae according to an endodiurnal rhythm (BÜNNING 1935a). MORI *et al.* reared *Drosophila* for about 240 generations in DD. The rhythm of emergence became obscure relative to the normal flies before the 135th generation, again becoming more distinct in the later generations. Mice are able to live without any external cycle (change of light and darkness, etc.) for several generations, and still exhibit the normal endodiurnal rhythm with the characteristic period (Fig. 19). BROWMAN kept rats in LL for 25 generations; despite this he found an approximate 24-hour rhythm in most individuals. Moreover, it was observed in plants that a specific period may be transferred through several generations to the offspring over many years (BÜNNING 1935b).

The hereditary nature of the circadian rhythms does not necessarily mean that the rhythms are manifested immediately after birth. In humans several weeks will pass after birth before the diurnal rhythm of the observed physiological activities becomes evident (*HELL-BRÜGGE; *HELLBRÜGGE *et al.*, see also Fig. 35). According to *RENSING, the embryonic stage of vertebrates in most, but not all, cases seems to be arrhythmic.

Of course, from observations such as these, we are not allowed to assume an absolute arrhythmicity of the organisms. Perhaps, we are

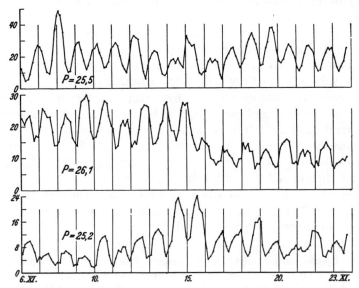

Fig. 19. Activity of mice, second generation in LL. P, average length of period, hours. After ASCHOFF 1955a

dealing only with phenomena of coupling and uncoupling of the observed physiological processes to the clock.

Genetic experiments. In plants the heredity of the endogenous diurnal rhythm was studied by producing hybrids of specimens with different periods (BÜNNING 1932). The first generation had intermediate periods. Although an unequivocal demonstration of Mendelian segregation was not possible, the original periods did appear in later generations. Experiments of this type must be repeated utilizing modern experimental facilities in which all conditions can be maintained as constant as necessary. In earlier days, conditions could not be kept constant enough, and thus the modifications of the period by environment interacted with the heredity differences to produce confusing results.

Another approach to genetic studies on the endogenous rhythm could be to crossbreed lines or races which differ in the critical day length of their photoperiodic responses. This critical day length can be determined by the endogenous rhythm (Chapter 13). DANILEVSKII, for example, found in the butterfly *Acronycta rumicis* that crossing different types of this insect yields intermediate values of the critical day length for the photoperiodic induction of diapause. This intermediate character was also maintained in the F_2-generation and after back-crossings with the F_1-generation. No clear segregation could be recognized, just as with the plants mentioned above. On the basis of these results, DANILEVSKII supposes that many genes participate in the determination of the critical day length.

KONOPKA and BENZER isolated mutants after treating male *Drosophila* with ethyl methane sulfonate (affecting the X-chromosome). They observed a mutant without rhythmicity, another with a period of 19 hours, and a third one with a 28-hour period. It is striking that the alteration of a single gene of the X-chromosome, as was the case here, can have these effects. Apparently, the mutation concerns the basic oscillator, since the effect can be detected both from the pupal eclosion rhythm and the adult activity rhythm.

BRUCE crossed wild-type strains of *Chlamydomonas reinhardi* with 24-hour and 21-hour periods of phototactic responsiveness. A single gene seems to confer the long-period character.

> Some investigators have attempted to influence the manifestations of endodiurnal activities by genetic manipulations (for instance, STADLER in *Neurospora*). Their findings, however, cannot necessarily be considered as contributions to a genetic analysis of the circadian clock, but perhaps only as contributions to the genetic analysis of the factors that are necessary for its manifestation, i.e., for coupling the specific physiological process to the clock. Moreover, it is doubtful, as was already pointed out in Chapter 2(b), that these rhythms may be looked upon as "genuine" circadian rhythms.

f. Loss of overt rhythmicity

Arctic regions. Some beetles of the arctic region showed only a very faint diurnal rhythm of running activity. Not even under marked experimental LD cycles did this become much stronger (Fig. 20). A

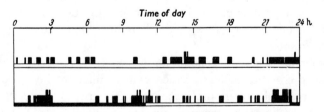

Fig. 20. Running activity of an individual *Carabus violaceus* L. from Lapland (latitude about 68°). Above, in the natural LL. Below, during experimental LD cycles. The height of the vertical black lines indicates the number of days on which the beetle was active at that particular time of day. Horizontal black lines represent darkness. Hours refer to Central European Time. After HEMPEL u. HEMPEL

rhythm of activity may also be missing in the penguin *Pygoscelis adeliae* (YEATES).

SWADE and PITTENDRIGH found several rodents from both arctic and temperate zones to have endogenous circadian rhythms of activity, when checked in the arctic and subarctic regions of Alaska. The mouse *Clethrionomys rufocanus* is active throughout 24 hours during the arctic summer, whereas it displays only nocturnal activity during the seasons with dark nights (PEIPONEN). Also, diurnal vertical migrations in certain planktonic species may be absent during the arctic summer (BOGOROV). Eskimos are reported not to show the normal 24-hour periodicity of urine production and potassium excretion, even when following a regular daily routine. In contrast, people from temperate regions, even when living several years in the arctic, continue to show the 24-hour periodicity of excretion (LOBBAN).

Differences in the intensity of the oscillations were found in plants as well. Soybean varieties that are unable to develop in areas distant from the equator (up to Scandinavia) are suitable for this study because of their photoperiodic response, which differs from the response of those varieties that are restricted to the vicinity of the equator. Apparently without exception, the soybean varieties flowering in long day conditions have an endodiurnal rhythm with smaller amplitudes (judged by their diurnal leaf movements) than those flowering only in tropical areas (Fig. 21). But several species of plants adapted to extreme arctic conditions (Spitsbergen, latitude 76–80°) still have endogenous circadian leaf movements (MAYER).

The absence of diurnal cycles in running activity under the conditions of continuous light found during the arctic summer does not

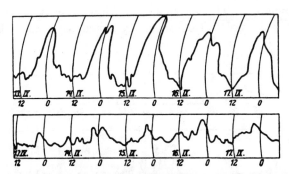

Fig. 21. Diurnal leaf movements in two varieties of soybean under the same laboratory conditions. Above: Otootan, an extreme short-day variety, flowering normally only in latitudes not too far from the tropics. Below: McRosties Mandarin, a nearly day-neutral variety, flowering also under long-day conditions. After BÜNNING 1948

indicate a failure of the clock. In certain cases at least, only the coupling of the locomotor activity to the clock is missing. Indeed, other processes, such as orientation by the sun, are coupled to the clock under the same conditions of the arctic summer (see Chapter 10). Especially, metabolic rhythms are more stable than rhythms of locomotor activity.

Whether a *synchronization* of the circadian rhythm is possible during the arctic summer is still another problem (Chapter 6).

Caves. A loss of the endogenous diurnal rhythm can be expected in organisms living in caves. PARK et al. (1941), could not find any rhythm of diurnal activity in the cave-crayfish *Orconectes pellucidus*, a blind species isolated from daily cycles for at least 25,000 generations. On the other hand, BROWN (1961) used the same data for calculations that still showed a faint diurnal rhythm.

More recent experiments by JEGLA and POULSEN confirmed this latter finding. Five of seven crayfishes tested under cave conditions (DD, 13°C) showed circadian rhythms of activity and O_2 consumption. These authors found clearer evidence for circadian rhythmicity in metabolic rate than in locomotor activity. As with organisms in arctic regions, the absence of an endogenous rhythm in locomotor activity may be the result of uncoupling activity from the clock or an effect of masking. BLUME et al. (see also GÜNZLER and similar results published by GINET) did not find any diurnal rhythm at all in the activity of an eyeless cave-crayfish (*Niphargus puteanus*) (Fig. 22). The mathematical analysis of the curves revealed that the periods ranged from 10 to 57 hours, even for a single specimen.

Wood beetles. Certain wood beetles do not show any diurnal rhythm of locomotor activity either (PARK 1935, 1937; PARK et al.; BECKER u. DAMASCHKE).

Domestication. The locomotor activity patterns of cockroaches offer classical examples of circadian rhythmicity. But under laboratory

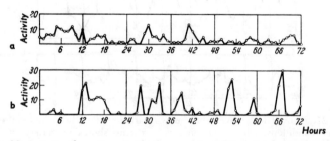

Fig. 22. Activity of a cave animal: the amphipod *Niphargus puteanus puteanus*. DD 8.5°C (a) without preceding LD cycles, (b) LD cycles prior to the record. After BLUME et al.

conditions a high percentage within a population of *Periplaneta americana* may display a random activity, even when kept in LD cycles. LIPTON and SUTHERLAND discuss the possibility that this is due to the limited survival value of such a rhythm under laboratory conditions.

General remarks. Adaptation to certain environmental conditions may result in a loss of overt rhythmicity. Most of these observations refer to rhythms in locomotor activity. In many cases at least, other circadian rhythms continue. Thus, we are apparently dealing with an uncoupling of locomotor activity from the clock.

g. Why do organisms use oscillations for chronometry?

In several cases, measuring time by oscillations certainly has advantages over using hour-glass processes. Oscillations allow "planning" not only for the next day, but for several days. However, in many cases of biochronometry, although using oscillations has no obvious advantage, it is not unusual to see organisms making use of them.

For reasons innate to the dynamics of the cell, "oscillations may be expected in biological systems whenever temporal constraints are required to seperate incompatible events in a process. Such oscillations will have frequencies which need bear no relation whatsoever to any environmental periodicity" (OATLEY and GOODWIN, p. 7). Thus, in order to develop a clock that allows an adaptation to the normal 24-hour periodicity, it was "only" necessary to *select* from the great variety of innate biochemical or biophysical oscillations. It was not necessary to *"construct"* special hour glasses.

Analogy with the evolution of the endogenous annual periodicity (see Chapter 13(e)) might favor this hypothesis. In those tropical regions where no adaptation to the seasons is necessary, biological cycles occur that deviate rather strongly from the 12-month cycle. Cycles in development and behavior lasting, for example, 2, 4, or 8 months, as well as 20 or 30 months, are known (for references see BÜNNING 1956, 1967). A selection from cycles such as these was the easiest starting point for temporal adaptation to regions with seasonal changes in the environment. No "construction" of hour-glass processes was necessary.

References to Chapter 2
(Endodiurnal Oscillations as the Principle of Many Physiological Time-Measuring Processes)

a. Reviews

*Aschoff, J., ed. 1965. *Circadian Clocks*. Amsterdam: Elsevier.—
*Aschoff, J. 1965. Circadian rhythms in man. *Science* 148:1427–1432.—
*Aschoff, J., K. Klotter, and R. Wever. 1965. Circadian Vocabulary. In *Aschoff (ed.), pp. X–XIX.
*Bruce, V. G. 1960. Environmental entrainment of circadian rhythms. In *Chovnick (ed.), pp. 29–48.—
———. Cell division rhythms and the circadian clock. In *Aschoff (ed.), pp. 125–138.—
*Bünning, E. 1958. *Tagesperiodische Bewegungen*. Encyclopedia Plant Physiol. XVII, 1, 579–656. Berlin-Göttingen-Heidelberg: Springer.—
———. Opening Address: Biological Clocks. In *Chovnick (ed.), pp. 1–9.
*Cameron, L. L., and G. M. Padilla, eds. 1966. *Cell Synchrony*. New York–London: Academic Press.—
*Caspers, H. 1957. Rhythmische Erscheinungen in der Fortpflanzung von *Clunio marinus* und das Problem der lunaren Periodizität bei Organismen. *Arch. Hydrobiol.* 18: Suppl. 415–594.—
*Chovnick, A., ed. 1960. Biological Clocks. *Cold Spring Harb. Symp. quant. Biol.* 25.—
*Colquhoun, W. P., ed. 1971. *Biological Rhythms and Human Performance*. London–New York: Academic Press.—
*Hastings, J. W., and A. Keynan. 1965. Molecular aspects of circadian systems. In *Aschoff (ed.), pp. 167–182.—
*Hastings, J. W., and B. M. Sweeney. 1959. The *Gonyaulax* clock. In *Photoperiodism and Related Phenomena in Plants and Animals*. R. B. Withrow, ed. Washington, D.C.: Am. Ass. Adv. Sci., pp. 567–584.
*Hellbrügge, T. 1960. The development of circadian rhythms in infants. In *Chovnick (ed.), pp. 311–323.—
*Hellbrügge, T., J. Ehrengut Lange, J. Rutenfranz, and K. Steht. 1964. Circadian periodicity of physiological functions in different stages of infancy and childhood. *Ann. N.Y. Acad. Sci.* 117:361–373.
*Jerebzoff, S. 1965. Manipulation of some oscillating systems in fungi by chemicals. In *Aschoff (ed.), pp. 183–189.—
*Jores, A. 1937. Die 24-Stunden-Periodik in der Biologie. *Tabul. Biol.* ('sGrav.) 14:77–109.
*Kalmus, H. 1935. Periodizität und Autochronie (= Ideochronie) als zeitregelnde Eigenschaften der Organismen. *Biol. Gen.* 11:93–114.—
*Kleitman, N. 1949. Biological rhythms and cycles. *Physiol. Rev.* 29: 1–30.—

———. 1963. *Sleep and Wakefulness*. Chicago: University Press.
°MITCHISON, J. M. 1971. *The Biology of the Cell Cycle*. Cambridge: Univ. Press.
°PADILLA, G. M., G. L. WHITSON, and I. L. CAMERON. 1969. *The Cell Cycle*. New York–London: Academic Press.—
°PARK, O. 1940. Nocturalism; the development of a problem. *Ecol. Monogr.* **10.**—
°PITTENDRIGH, C. S. 1960. Circadian rhythms and the circadian organization of living systems. In °CHOVNICK (ed.), pp. 159–184.—
°PITTENDRIGH, C. S., and V. G. BRUCE. 1957. An oscillator model for biological clocks. In *Rhythmic and Synthetic Processes in Growth*. D. Rudnick, ed. Princeton: Univ. Press 1957. Pp. 75–109.
°REMMERT, H. 1962. *Der Schlüpfrhythmus der Insekten*. Wiesbaden: Franz Steiner.—
°RENSING, L. 1965. Circadian rhythms in the course of ontogeny. In °ASCHOFF (ed.), pp. 399–405.—
°RICHTER, C. P. 1965. *Biological Clocks in Medicine and Psychiatry*. Springfield, Ill.: Charles C Thomas.
°SWEENEY, B. 1960. The photosynthetic rhythm in single cells of Gonyaulax polyedra. In °CHOVNICK (ed.), pp. 145–148;—
———. 1965. Rhythmicity in the biochemistry of photosynthesis in Gonyaulax. In °ASCHOFF (ed.), pp. 190–194;—
———. 1969. *Rhythmic Phenomena in Plants*. London–New York: Academic Press.
°WELSH, J. H. 1938. Diurnal rhythms. *Quart. Rev. Biol.* **13**, 123–139.—
°WILKINS, M. B. 1960. The effect of light upon plant rhythms. In °CHOVNICK (ed.), pp. 115–129.
°ZEUTHEN, E., ed. 1964. *Synchrony in Cell Division and Growth*. New York: Interscience Publishers.

b. Other references

ASCHOFF, J. 1955a. *Pflügers Arch. ges. Physiol.* **262**:51–59;—
———. 1955b. *Klin. Wschr.* 545–551;—
———. 1966. *Ecology* **47**:657–662.—
ASCHOFF, J., u. K. HONMA. 1959. *Z. vergl. Physiol.* **42**:383–392.—
ASCHOFF, J., u. J. MEYER-LOHMANN. 1954. *Pflügers Arch. ges. Physiol.* **260**: 170–176.—
ASCHOFF, J., u. R. WEVER. 1962. *Naturwiss.* **49**:337–342.—
AUSTIN, B. 1968. *Ann. Bot.* **32**:261–278.
BALL, N. G., I. J. DYKE, and M. B. WILKINS. 1957. *J. exp. Bot.*, **8**:339–347.—
BARANETZKY, J. 1879. *Mém. Acad. Sci. St. Pétersbourg*, VII Sér. **27**:1–91.—
BECKER, G., u. K. DAMASCHKE. 1963. *Z. ang. Entomol.* **51**:323–334.—
BELING, I. 1929. *Z. vergl. Physiol.* **9**:259–338.—
BENNETT, M. F., and C. B. GUILFORD. 1971. *Z. vergl. Physiol.* **74**:32–38.—

BERLINER, M. D., and P. W. NEURATH. 1965a. *J. cell. a. comp. Physiol.* **65**: 183–193;–
———. 1965b. *Mycologia* **57**:809–817.–
BIANCHI, D. E. 1964. *J. Gen. Microbiol.* **35**:437–445.–
BLUME, J., E. BÜNNING, u. E. GÜNZLER. 1962. *Naturwiss.* **49**:525.
BOGOROV, B. G. 1946. *J. mar. Res.* **6**:25–32.–
BOURRET, J. A., R. G. LINCOLN, and B. H. CARPENTER. 1969. *Science* **166**: 763–764.–
BREMER, H. 1926. *Z. wiss. Insektenbiol.* **21**:209–216.–
BRETZL, H. 1903. *Botanische Forschungen des Alexanderzuges.* Leipzig: B. G. Teubner.–
BROWMAN, L. G. 1952. *Amer. J. Physiol.* **168**:694–697.–
BROWN, F. A. 1961. *Nature (Lond.)* **191**:929–930.–
BRUCE, V. G. 1972. *Genetics* **70**:537–548.–
BUCK, J. B. 1937. *Physiol. Zool.* **X**:45–58.–
BÜHLER, A., u. E. BÜNNING. 1965. *Arch. Mikrobiol.* **52**:80–82.–
BÜHNEMANN, F. 1955. *Biol. Zbl.* **74**:1–54.–
BÜNNING, E. 1932. *Jahrb. wiss. Bot.* **77**:283–320;–
———. 1935a. *Ber. dtsch. Bot. Ges.* **53**:594–623;–
———. 1935b. *Jahrb. wiss. Bot.* **81**:411–418;–
———. 1948. *Z. Naturforsch.* **3b**:457–464;–
———. 1956. *Encyclop. Plant Physiol.* **2**:878–907;–
———. 1967. *Ann. N.Y. Acad. Sci.* **138**:2515–524.–
BÜNNING, E., u. K. STERN. 1930. *Ber. dtsch. Bot. Ges.* **48**:227–252.–
BÜNNING, E., u. M. TAZAWA. 1957. *Planta (Berl.)* **50**:107–121.–
BÜNSOW, R. 1953. *Planta (Berl.)* **42**:220–253.
CLAUSER, G. 1954. *Die Kopfuhr.* Stuttgart: F. Enke.–
CLAUSS, H., u. B. SCHWEMMLE. 1959. *Z. Bot.* **47**:226–250.–
COLQUHOUN, W. P. 1971. In *Biological Rhythms and Human Performance.* W. P. COLQUHOUN, ed. London–New York: Academic Press.–
DANILEVSKII, A. S. 1957. *Entomol. Obozrenic* **36**:5–27.–
DECOURSEY, P. J. 1960. *Cold Spring Harbor Symp. quant. Biol.* **25**:49–55;–
———. 1961. *Z. vergl. Physiol.* **44**:331–354.–
DE MAIRAN. 1729. *Observation botanique.* Histoire de l'Academie Royale des Sciences Paris, p. 35.
DEMOLL, R. 1911. *Zool. Jb. Physiol.* **30**:159–180.–
EDMUNDS, L. N. 1965. *J. cell. a. comp. Physiol.* **66**:147–181;–
———. 1966. *J. cell. a. comp. Physiol.* **67**:35–43;–
———. 1971. In *Biochronometry.* M. MENAKER, ed. Washington, D.C.: *Nat. Acad. Sci.* Pp. 594–611.–
ESKIN, A. 1971. In *Biochronometry.* M. MENAKER, ed. Washington, D.C.: *Nat. Acad. Sci.* Pp. 55–80.–
FOREL, A. 1910. *Das Sinnesleben der Insekten.* München: Reinhardt.–
FORSGREN, E. 1928. *Scand. Arch. Physiol.* **53**:137;–
———. 1935. *Über die Rhythmik der Leberfunktion, des Stoffwechsels und des Schlafes.* Stockholm.–

FUNCK, H. 1960. *Die renale Ausscheidung von Natrium bei Neugeborenen und Säuglingen.* Diss. München.
GEISLER, M. 1961. *Z. Tierpsychol.* **18**:389–420.–
GERRITZEN, F. 1955. *Acta med. scand. Suppl.* **307**:150–152.–
GINET, R. 1960. *Ann. de Spéléologie* **15**:1–254.–
GÜNZLER, E. 1964. *Biol. Zbl.* **83**:677–694.
HALBERG, F., S. SIFFRE, M. ENGELI, D. HILLMAN, et A. REINBERG. 1965. *C. R. Acad. Sci. Paris* **260**:1259–1262.–
HEMMINGSEN, A. M., and N. B. KRARUP. 1937. *Kgl. Dansk. Vidensk. Selskab. Biol. Medd.* **13**(7):1–61.–
HEMPEL, G., u. I. HEMPEL. 1959. *Naturwiss.* **42**:77–78.–
HOFFMANN, K. 1955. *Z. vergl. Physiol.* **37**:253–262.–
———. 1957. *Naturwiss.* **44**:359–360.–
———. 1959. *Z. vergl. Physiol.* **42**:422–432.–
HORSTMANN, C. 1935. *Biol. Zentralbl.* **55**:93–97.–
JEGLA, TH. C., and TH. L. POULSON. 1968. *J. exp. Zool.* **168**:273–282.–
JEREBZOFF, S. 1961. Etude de phénomènes périodiques provoqués par des facteurs physiques et chimiques chez quelques champignons. Thèses Univ. Toulouse.–
JOHNSON, M. S. 1926. *J. Mammal.* **7**:245–277.–
JONES, M. B., and T. A. MANSFIELD. 1970. *J. exp. Bot.* **21**:159–163.–
KALMUS, H. 1934. *Z. vergl. Physiol.* **20**:405–419.–
KARAKASHIAN, M. W. 1965. In *Circadian Clocks.* J. ASCHOFF, ed. Amsterdam. Pp. 301–304.–
KIESEL, A. 1894. *S. B. Akad. Wiss. Wien* **103**:97–139.–
KLEINHOONTE, A. 1929. *Arch. Néerl. Sci. ex. et nat. III b* **5**:1–110.–
———. 1932. *Jahrb. wiss. Bot.* **75**:679–725.–
KLEITMAN, N. 1940. *Biol. Bull.* **78**:403–411.–
KONOPKA, R. J., and S. BENZER. 1971. *Proc. Nat. Acad. Sci. U.S.A.* **68**:2112–2116.
LEINWEBER, F. J. 1956. *Z. Bot.* **44**:337–364.–
LEVENGOOD, M. C. 1969. *Z. vergl. Physiol.* **62**:153–166.–
LIPTON, G. R., and D. J. SUTHERLAND. 1970. *J. Insect Physiol.* **16**:1555–1566.–
LOBBAN, M. C. 1967. *Quart. J. exp. Physiol. Cog. Med. Sci.* **52**:401–410.–
LOHMANN, M. 1967. *Biol. Zentralbl.* **86**:623–628.–
LOWE, CH. H., D. S. HINDS, P. J. LARDNER, and K. E. JUSTICE. 1967. *Science* **156**:531–534.–
LUTZ, F. E. 1932. *Amer. Mus. Novitates* 550.
MAYER, W. 1966. *Planta (Berl.)* **70**:237–256.–
MENAKER, M. 1961. *J. cell a. comp. Physiol.* **57**:81–86.–
MIYATA, H. 1970. *Plant a. Cell Physiol.* **11**:293–301.–
MORI, S., S. YANAGISHIMA, and N. SUZUKI. 1966. *Biometeorology II.* Proc. 3rd Int. Biometeor. Congr. Pp. 550–563. Oxford: Pergamon Press.
NOWOSIELSKI, J. W., and R. L. PATTON. 1963. *J. Insect Physiol.* **9**:401–410.

OATLEY, K., and B. C. GOODWIN. 1971. In *Biological Rhythms and Human Performance*. W. P. COLQUHOUN, ed. London–New York: Academic Press.–

PALMER, J. D. 1964. *Comp. Biol. Physiol.* **12**:273–283.–

———, L. LIVINGSTON, and F. D. ZUSY. 1964. *Nature* **203**:1087–1088.–

PARK, O. 1935. *Ecology* **16**:152–163.–

———. 1937. *J. Animal Ecol.* **6**:239–253.–

———, and J. G. KELLER. 1932. *Ecology* **13**:335–346.–

———, T. W. ROBERTS, and S. J. HARRIS. 1941. *Amer. Naturalist* **45**:154–171.–

PEIPONEN, V. A. 1962. *Arch. Soc. Zool. Bot. Fennica "Vanamo"* **17**:171–178.–

PFEFFER, W. 1873. *Physiologische Untersuchungen*. Leipzig.–

———. 1875. *Die periodischen Bewegungen der Blattorgane*. Leipzig.–

———. 1911. *Abh. Math. Phys. Kl. Kgl. Sächs. Ges. Wiss.* **32**:163–295.–

———. 1915. *Abh. Math. Phys. Kl. Kgl. Sächs. Ges. Wiss.* **34**:1–154.

QUINKE, H. 1893. *Arch. exp. Path.* **32**:211–240.

REINBERG, A., F. HALBERG, J. GHATA, et M. SIFFRE. 1966. *C. R. Acad. Sci. Paris* **262**:782–785.–

RENSING, L. 1968. *Verh. Deut. Zool. Ges. Leipzig*. Acad. Verl. Ges. Pp. 298–307.–

RICHTER, C. P. 1922. A behavioristic study of the activity of the rat. *Comp. Physiol. Monogr.* **1**.–

ROBERTS, S. K. 1959. Ph.D. Thesis. Princeton.–

SACHS, J. 1857. *Bot. Ztg.* **15**(47):809–815.–

———. 1863. *Flora* **30**:465–472.–

SARGENT, M. L., W. R. BRIGGS, and D. O. WOODWARD. 1969. *Plant Physiol.* **41**:1343–1349.–

SCHMIDLE, A. 1951. *Arch. Mikrobiol.* **16**:80–100.–

SEMON, R. 1905. *Biol. Zentralbl.* **25**:241–252.–

———. 1908. *Biol. Zentralbl.* **28**:225–243.–

STADLER, D. R. 1959. *Nature (Lond.)* **184**:169–171.–

STOPPEL, R. 1916. *Z. Bot.* **8**:609–684;–

———. 1938. *Ber. dtsch. Bot. Ges.* **66**:177–190;–

———. 1940. *Planta (Berl.)* **30**:695–715.–

SUSSMAN, A. S., R. W. LOWRY, and F. DURKEE. 1964. *Am. J. Bot.* **51**:243–252.–

SWADE, R. H., and C. S. PITTENDRIGH. 1967. *Amer. Naturalist* **101**:431–466.

TWEEDY, D. G., and W. P. STEPHEN. 1971. *Comp. Bioch. Physiol.* **38**:213–231.

UEBELMESSER, E. R. 1954. *Arch. Mikrobiol.* **20**:1–33.–

USHATINSKAYA, R. S. 1969. In *Periodicity in Insects' Individual Development* (in Russian). K. W. ARNOLDY, ed. Nauka Acad. Sci. USSR. Moskau. pp. 13–98.

WAHL, O. 1932. *Z. vergl. Physiol.* **16**:529–589.–

WALKEY, D. G. S., and R. HARVEY. 1967. *Trans. Brit. Mycol. Sco.* **50**:229–240; 241–249.–
WARREN, D. M., and M. B. WILKINS. 1961. *Nature (Lond.)* **101**:686–688.–
WILKINS, M. B. 1959. *J. exp. Bot.* **10**:377–390;–
———. 1967. *Planta (Berl.)* **72**:66–77.–
WOLF, J. 1958. From E. BÜNNING: Encyclop. Plant Physiol. XII, 2, S.595.
YEATES, G. W. 1971. *J. natur. Hist.* **5**:103–112.
ZINN, J. G. 1759. *Hamburg, Magazin* **22**:40–50.

3. Periodicity Fade-Out; Initiation by External Factors

... könne man schliessen, die Bewegung des Einschlafens und Erwachens hinge mit einer, dem Gewächse einwohnenden Anlage zu periodischer Bewegung zusammen; letztere würde aber wesentlich durch die erregende Einwirkung des Lichtes in Tätigkeit gesetzt. ...

DE CANDOLLE, Pflanzen-Physiologie, Bd. 2, pp. 639–640, 1835 (Original: Paris 1832).

(The periodicity of leaf movements is inherited, but light is necessary to evoke it.)

a. Necessity of impulse

It was mentioned earlier that animals and plants can show an endogenous diurnal rhythm even if they have not been exposed to diurnal cycles of environmental factors during their embryonic development. Very often these organisms do not show any periodicity at first, but a single stimulus is then sufficient to initiate the rhythm (BÜNNING 1931, 1935a, b; BALL and DYKE; Fig. 23, 24). Since the necessary single stimulus can be compared with the stimulus required for the reinitiation of a rhythm which has faded out, the phenomenon of a rhythm fading out will be discussed first.

b. Fade-out of the periodicity

Duration of free-running oscillations. A rhythm that has been initiated and controlled by LD cycles can be observed under constant

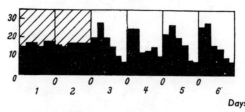

Fig. 23. *Drosophila*. Initiation of periodic emergence by the transition from DD (shaded) to LL. For details compare with Fig. 8. After BÜNNING 1935a

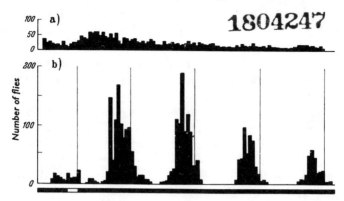

Fig. 24. *Drosophila*. Initiation of an eclosion rhythm by exposing a dark-grown culture to 4 hours of white light. (a) control in DD. (b) light period at indicated time. The vertical guide lines are at 4, 24, 48, 72, 96, and 120 hours after the start of the light stimulus. The ordinates represent numbers of eclosed flies per hour. After *Pittendrigh and Bruce

conditions for different lengths of time—often only for a few days, in some cases for one to several weeks (Figs. 25, 27), or even for several months. Good examples of rhythms that persist over long periods are the diurnal pigment migrations in the fiddler crab *Uca* (Brown and Webb), the diurnal fluctuations of sporulation in the alga *Oedogonium*, and many of the activity cycles in rodents. The continuation of these activity cycles for several months under constant conditions has often been described, justifying the term of "self-sustained oscillations" (*Pittendrigh). It has been known for many years that the diurnal leaf movements of *Phaseolus* and other plants may also continue for several weeks or longer than one month under constant conditions.

Long lasting oscillations may even persist during hibernation, as

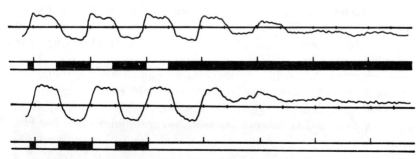

Fig. 25. *Chenopodium amaranticolor*. Diurnal leaf movements. Dark periods black, light periods white. Vertical lines indicate 24-hour periods. In DD or in LL, the movements continue only for a short period. After Könitz (original)

has been shown for the bat *Myotis myotis* (POHL 1961), as well as for several rodents (SAINT GIRONS; KAYSER). The self-sustaining character of the persistent rhythms becomes especially clear from the deviations from the exact 24-hour period. In the rodent *Glis glis,* POHL (1965) measured rhythms of spontaneous locomotor activity with periods of 26.5 hours during hibernation.

Fade-out of overt rhythmicity does not necessarily mean that the cells or tissues involved no longer have circadian oscillations. If, for example, a cell's ability for changes in turgor pressure is lost, this cell of course can no longer show overt circadian cycles of processes that are dependent on these changes. This is the reason for the damping and fading away of circadian leaf movements.

Oscillations may also continue through different stages of morphogenesis. In *Drosophila,* for example, rhythmicity continues from the larval stages up to the emerging of flies from the pupae (BÜNNING 1935a; similar observations by BRETT; SCHERER; MINIS and PITTENDRIGH; for further discussion see PITTENDRIGH and SKOPIC).

How long such an oscillation can continue is dependent not only on the individual object but even more on the given conditions. Quite often the fade-out is quicker in LL than in DD. In plants the light quality also has an important influence. In beans the leaf movement will fade out faster in far-red light than in DD, and the movement continues much longer in red light than in DD. For that reason the movement is now recorded mostly in LL of fluorescent tubes, which produce the favorable red light but do not emit the unfavorable far-red light, which dominates in incandescent bulbs. The temperature level is also important, as is the physiological condition, which in turn depends upon the previous treatment of the object.

Fade-out by desynchronization. In some cases this fade-out is evidently due to a gradual desynchronization, i.e., to phase-shifts of different magnitude in the different organs or even in the individual cells of the test organisms. This can be observed in the organs of some plants where the different parts, either individual cells or groups of cells, evidently become independent of one another (Fig. 26). The recorded curves will then show a disintegration into several part-oscillations which can be explained as superimposition, as indicated by the shape of the peaks. This became quite evident during the observations by TODT in *Cichorium intybus.* Based on the rhythm of flower-opening in LL, one can say that the synchronization is lost first among the individual plants, then among the different composite flowers of a single plant, then among the individual flowers of a composite flower, and finally among the petals of an individual flower (Fig. 28). Several weeks of continuous light are required to obtain such a complete desynchronization. Experiments with zoological objects which permit this kind of interpretation have also been described.

If the periodicity in higher animals functioned throughout only by central control, and if the control centers were not capable of any

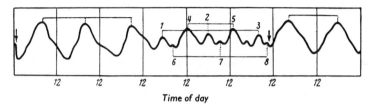

Fig. 26. *Phaseolus coccineus.* Leaf movements in DD. Desynchronization by a light stimulus at time indicated by arrow. A second light stimulus (second arrow) causes new synchronization. It may be clearly seen from the shape of the curves that the maxima at 1, 2, and 3, or at 4 and 5, or 6, 7, and 8 belong to a partial oscillation of an approximately 24-hour rhythm. After BÜNNING 1935b

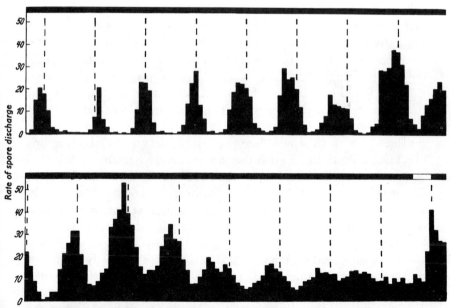

Fig. 27. *Daldinia concentrica.* Spore discharge in continuous darkness. The horizontal strip above indicates conditions of illumination, black representing darkness and white representing light. Vertical interrupted lines give position of midnight. Eventual damping-out in the course of approximately two weeks. On the last day the rhythm was re-induced by a new light period. The continuation of this experiment is not shown here. After INGOLD and COX

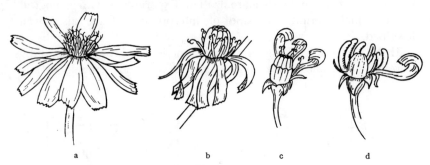

Fig. 28. Inflorescences of *Cichorium intybus*. (a) grown in normal LD cycles, (b) for 16 days, (c) for 22 days, (d) for 28 days in LL. After TODT

desynchronization, then the animal as a whole should always display periodic behavior, assuming that fading out is due only to desynchronization. Yet this is not the actual behavior of higher animals. HALBERG et al. (1953) found that after nine days of LL one can no longer distinguish an eosinophil rhythm in the blood of mice (in DD it can still be recognized after thirteen days). Of course, it is conceivable that desynchronizations will occur in the controlling glands, so that the individual cells will oscillate in shifted phases after some time. There are indications that, in vertebrate animals, organs usually working in cycles might become desynchronized in this way. For instance, EBBECKE found in kidneys a rhythmical alternation of glomerulus and tubulus activity (associated with an alternation of concentrating and filtrating activity). Apparently, either all the nephronic groups may synchronically work together, or the phases in the individual cell groups of a kidney are shifted so that they are no longer synchronized.

In some mammals a desynchronization of rhythms was found in LL or DD, although the individual rhythms still continued and could be clearly distinguished. This was reported by HALBERG and BARNUM on rhythms of metabolism and mitotic activity in certain organs of the mouse. Later on we shall give some more examples of the desynchronization of certain physiological activities in the body of higher organisms, but this does not permit the conclusion that the different activities function aperiodically.

Fade-out in the individual cell. Many observations indicate the possibility that rhythms also fade out in the individual cell. Observations on *Gonyaulax* are convincing: in LL the diurnal periodicity of photosynthetic capacity can no longer be shown even in individual cells. The absence of this rhythm in populations kept in LL is really the expression of a periodicity fade-out in the individual cell and not the result of a desynchronization (SWEENEY).

The absence of a rhythm in single cells of higher plants can be deduced from experiments of WASSERMANN. As long as the plants (*Phaseolus*) are exposed to LD cycles, the volume of the nuclei fluctuates diurnally. These fluctuations will also continue for a certain time under constant conditions. Yet no volume fluctuations occur when the plants are cultivated under continuous light. Especially striking is the fact that the extremely large volumes cannot be found at all under these conditions. Only after exposing the plants to dark periods which again initiate the rhythm, displayed in the leaf movements, can one observe the fluctuations of the nuclei. Only then can large volumes be recognized again.

In the giant cells of the alga *Acetabularia*, the rhythms in photosynthetic capacity and in chloroplast shape are lost after three weeks in DD (VANDEN DRIESSCHE).

These observations permit the conclusion that when a rhythm in individual cells fails, certain extreme physiological conditions cease to be attainable. This is one of the reasons why organisms may be damaged in LL (see Chapter 14).

Fade-out following light perturbations. Light perturbations normally cause phase shifts (Chapter 6). But under certain conditions, a light perturbation of appropriate strength, offered during a certain phase of the circadian cycle, can result in stopping the clock—i.e., it can lead to an arrhythmic status. This has been shown both for animals (WINFREE) and for plants (TAKIMOTO and HAMNER).

c. The nature of the initiating stimuli

There are various ways in which a rhythm can be initiated after the fade-out of a previous rhythm or in the absence of a rhythm, when the organism has been brought up under constant conditions. In some cases it is sufficient to interrupt the DD by a short light stimulus. Interrupting LL by a dark period can have the same effect, as can the change from DD to LL (Figs. 23, 29, and 31) or from LL to DD, or even a change in light intensity (Fig. 30). A change in temperature can also be sufficient for an initiation.

A single dark period of a few hours, interrupting LL, will initiate the rhythm in several plant species (Fig. 31). In some cases about 4 hours are enough, in other cases it takes 8–12 hours (examples are given by WASSERMANN; ISAAC and ABRAHAM; RUDDAT). In *Paramecium* a single dark period of a few hours may initiate the circadian rhythm

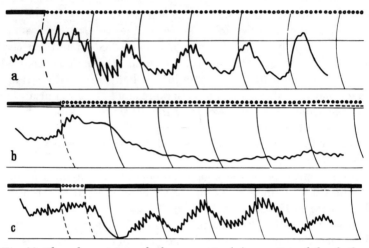

Fig. 29. *Phaseolus coccineus,* leaf movements. (a) evocation of the rhythm by transition from DD to LL (red). (b) no rhythm when far-red is offered in addition to red. (c) evocation by a short period of red. Broad black lines indicate darkness, circles red light, broken line far-red. Guide lines 24 hours apart. After LÖRCHER

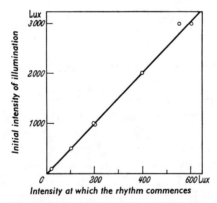

Fig. 30. *Bryophyllum fedtschenkoi.* Rhythm of CO_2 output. The figure shows the linear relation between the initial intensity of illumination to which the leaves were subjected and the intensity at which the rhythm commenced. After °WILKINS 1960a

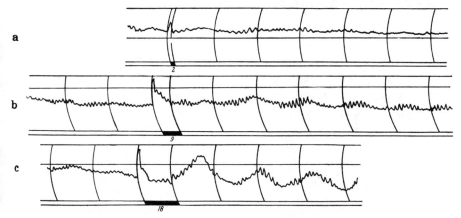

Fig. 31. *Phaseolus coccineus*, leaf movements. The plants were grown from seed germination in LL. Behavior after single dark periods of 2, 9, and 18 hours. Guide lines 24 hours apart. After WASSERMANN

of cell division (VOLM). If the periodicity can be initiated by a single light stimulus, minutes or even fractions of a second may be sufficient (*BRUCE).

d. Special questions about initiation by light and darkness

Absence of a recognizable periodicity can be due to a lack of synchronization of the individual cells. In this case the initiation synchronizes the independently oscillating cells by determining the phase position. Phase-shifting by light is brought about through processes we shall discuss later. If the periodicity has faded out in the individual cells, the initiation represents an entirely different process. The action spectra indicate that pigment systems other than those mediating the shifting of phases by light were responsible. This can be concluded from experiments with beans: The initiation of a periodicity which had faded out is possible only by red light, whereas phase-shifting can also be achieved with other light qualities (Fig. 29). The special significance of red light for initiating the periodicity is also evident in the experiments of BALL and DYKE with *Avena* coleoptiles. On the other hand, the action spectrum with *Kalanchoe* flowers indicated that there the initiation is due to absorption by chlorophyll (KARVE et al.). Initiation in fungi that respond to light is possible only with blue light. Since various objects respond so differently, we can assume that the impulse for oscillation can be given by some totally different processes within the cell caused by light absorption in various pigment systems.

In LL an increase of intensity can also act as an initiation. The initiating effect of a dark period can be reproduced by decreasing the light intensity. For example, in *Bryophyllum* it is necessary to decrease the light intensity by 80% in order to initiate the rhythm of periodical CO_2 output (Fig. 31). Approximately the same relation was found in the luminescence of the unicellular alga *Gonyaulax*.

Initiation by transition stimuli of the type mentioned above usually results in a phase position such as might be predicted on the basis of the phase position in the normal 24-hour LD cycle. Besides the time of transition from darkness to light, or from light to darkness, it is also important to consider the length of the dark period interrupting continuous light. The study of these relations may aid in understanding the processes involved. For example, in beans (*Phaseolus coccineus*) grown in LL, it takes a dark period of from 9 to 10 hours to initiate periodicity (Fig. 31); similar results concerning growth rhythms in the broad bean, *Vicia faba*, were published by MILLET. If the dark period is shorter, it cannot become effective even if it is supplemented with another short dark period which follows some hours later (Fig. 32). Thus, the process going on in the dark diminishes during a subsequent light period. This gives the impression that dark periods of less than ten hours only partially "tighten" or "release" the oscillating system, and after the end of the dark period the changes will gradually wear off again. Only if the process can be extended to a critical value, which takes 9 to 10 hours of darkness, will it exceed the threshold and thereby give the impulse for the oscillation (see Chapter 7).

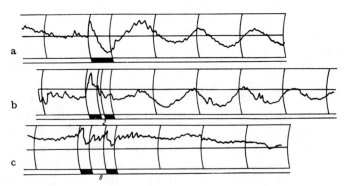

Fig. 32. Similar to Fig. 31, but with initiation by a dark period of 12 hours. The dark period was interrupted by (b) 2 hours or (c) 8 hours of light. After WASSERMANN

e. Initiation by low temperature

Several hours of low temperature (e.g., 5°C) without change in light intensity exert an effect similar to that of a dark period interrupting LL. This phenomenon, which also supports the interpretation of the inductive dark period given above, was observed in beans (initiating leaf movements, see WASSERMANN), in the alga *Oedogonium* (initiating the sporulation rhythm, see RUDDAT), and in *Paramecium* (initiating division rhythm, see VOLM). In order to be effective, the low temperatures must be applied for several hours, comparable to the length of time required when a dark period interrupts continuous light.

f. Absence of rhythms in early developmental stages

Some plants cultivated in LL or in DD show no overt rhythm. In several vertebrates the endogenous rhythm becomes evident immediately after birth, particularly with respect to running activity. Young chickens show it immediately after their emergence from the egg. Yet birth by itself may well represent a synchronizing stimulus. In mammals, of course, the environmental conditions of the embryo might have a synchronizing effect even before birth.

Nevertheless, many physiological functions of mammals seem to be arrhythmic during the embryonic stages. But there are also indications that certain processes are already rhythmic during these stages (for references see *HARKER; *RENSING). In birds observations have been made which indicate the absence of synchronization in embryos. In young chicken embryos, it was found that rhythmical changes occur in the glycogen content of the liver. These have a striking resemblance to desynchronization in plants: the rhythm is resolved into partial oscillations (Fig. 33). Yet young chickens show a diurnal rhythm in their running activity immediately after birth. (ASCHOFF u. MEYER-LOHMANN). A diurnal cycle in the functions of their liver becomes evident after emerging from the egg (Fig. 34). One might assume that the emergence from the egg is a synchronizing factor.

But even if the organism is exposed to diurnal LD cycles or to other diurnal cycles, it is still possible that the physiological rhythm does not become evident for some time. The gradual development of

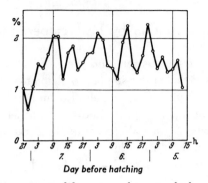

Fig. 33. Glycogen content of liver in embryonic chickens from 7 to 5 days before hatching. Average values from 12–16 embryos per point. Abscissa: times of day and days before hatching. Ordinate: liver glycogen, %. After PETREN, simplified

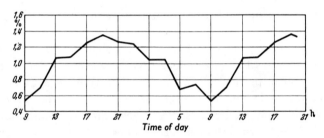

Fig. 34. As in Fig. 33, but in 5-day-old chickens in LL. Values from about 15 animals. After PETREN, simplified

the normal rhythm of humans in being awake and asleep should be mentioned again (*HELLBRÜGGE; HELLBRÜGGE et al.; KLEITMAN and ENGELMANN). This diurnal rhythm in sleeping habits becomes evident only after the sixth week of life (Fig. 35). Similar observations were made on pulse frequency, water excretion, and other functions which eventually develop a diurnal periodicity.

However, this does not mean that organisms in their young stage are without any rhythm. In order to recognize a periodicity in physiological activities, it is necessary for these processes to be linked to the internal clock. Observations indicate that whether a physiological process is coupled to the clock or not may depend on the given conditions within the body. The situation is similar with respect to ecological adaptations (see also Chapter 2(f)).

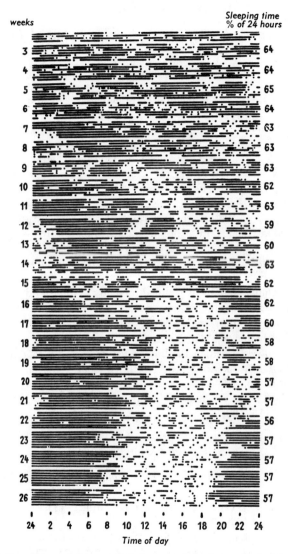

Fig. 35. Frequency of sleep, wakefulness, and feeding periods of a child from the 11th to 182nd day of life. Each line of the abscissa represents a 24-hour calendar day. The solid lines represent sleep, the intervals in between are periods of wakefulness and the dots represent feeding periods. There is a striking preponderance of the periods of wakefulness between 8 A.M. and 8 P.M. starting with the 16th week after birth. After KLEITMAN and ENGELMANN

References to Chapter 3
(Periodicity Fade-Out; Initiation by External Factors)

a. Reviews

*BRUCE, V. G. 1960. Environmental entrainment of circadian rhythms. In *CHOVNICK (ed.), pp. 29–48.
*CHOVNICK, A., ed. 1960. Biological Clocks. Cold Spring Harb. Symp. quant. Biol. **25**.
*HARKER, J. E. 1964. The Physiology of Diurnal Rhythms. Cambridge: Univ. Press.–
*HELLBRÜGGE, TH. 1960. The development of circadian rhythms in infants. In *CHOVNICK (ed.), pp. 311–323.
*PITTENDRIGH, C. S. 1960. Circadian rhythms and the circadian organization of living systems. In *CHOVNICK (ed.), pp. 159–184.–
——— and V. G. BRUCE. 1957. An oscillator model for biological clocks. In Rhythmic and Synthetic Processes in Growth. D. RUDNICK, ed. Princeton Univ. Press. Pp. 75–109;–
———. 1959. Daily rhythms as coupled oscillator systems and their relation to thermoperiodism and photoperiodism. In Photoperiodism and Related Phenomena in Plants and Animals. R. B. WITHROW, ed. Washington: Am. Ass. Adv. Sci. Pp. 475-505.
*RENSING, L. 1965. Circadian rhythms in the course of ontogeny. In Circadian Clocks. J. ASCHOFF, ed. North Holland Publishing Company: Amsterdam. Pp. 399–404.
*WILKINS, M. B. The effect of light upon plant rhythms. In *CHOVNICK (ed.), pp. 115–129.

b. Other references

ASCHOFF, J. u. J. MEYER-LOHMANN. 1954. Pflügers Arch. ges. Physiol. **260**: 170–176.
BALL, N. G., and I. J. DYKE. 1954. J. exp. Bot. **5**:421–433;–
———. 1956. J. exp. Bot. **7**:25–41.–
BRETT, W. J. 1955. Ann. Entomol. Soc. Amer. **48**:119–131.–
BROWN, F. A., and H. M. WEBB. 1948. Physiol. Zool. **21**:371–381.–
BÜNNING, E. 1931. Jahrb. wiss. Bot. **75**:439–480;–
———. 1935a. Ber. dtsch. Bot. Ges. **53**:594–623;–
———. 1935b. Jahrb. wiss. Bot. **81**:411–418.
EBBECKE, U. 1931. Pflügers Arch. ges. Physiol. **226**:761–773.
HALBERG, F., and C. P. BARNUM. 1961. Amer. J. Physiol. **201**:227–230.–
———, M. B. VISSCHER, and J. J. BITTNER. 1953. Amer. J. Physiol. **174**:109–122.–

HELLBRÜGGE, TH., J. LANGE, u. J. RUTENFRANZ. 1959. *Beitr. Arch. Kinderheilk.* **39.**
INGOLD, C. T., and V. J. COX. 1955. *Ann. Bot.* **19:**201–209.
ISAAC, I., and G. H. ABRAHAM. 1959. *Can. J. Bot.* **37:**801–814.
KARVE, A., W. ENGELMANN, u. G. SCHOSER. 1961. *Planta (Berl.)* **56:**700–711.–
KAYSER, CH. 1965. *Arch. Sci. physiol.* **19:**369–413.–
KLEITMAN, N., and TH. G. ENGELMANN. 1953. *J. appl. Physiol.* **6:**269–282.–
LÖRCHER, L. 1958. *Z. Bot.* **46:**209–241.
MILLET, B. 1970. *Thése Fac. Sc. de l'Univ. de Besançon.*–
MINIS, D. H., and C. S. PITTENDRIGH. 1968. *Science* **159:**534–536.
PETREN, T. 1955. Suppl. Nr. 307 der *Acta med. scand.* 42–47.–
PITTENDRIGH, C. S., and S. D. SKOPIC. 1970. *Proc. Nat. Acad. Sci. USA* **65:**500–507.–
POHL, H. 1961. *Z. vergl. Physiol.* **45:**109–153;–
———. 1965. *Naturwiss.* **52:**269.
RUDDAT, M. 1960. *Z. Bot.* **49:**23–46.
SAINT GIRONS, M. C. 1965. In *Circadian Clocks.* J. ASCHOFF, ed. Amsterdam: North Holland Publishing Company. Pp. 321–323.–
SCHERER, L. E. 1965. *Florida Entomol.* **47:**227–233.–
SWEENEY, B. 1960. In °CHOVNICK (ed.), pp. 145–148.
TAKIMOTO, A., and K. C. HAMNER. 1965. *Plant Physiol.* **40:**855–858.–
TODT, D. 1962. *Z. Bot.* **50:**1–21.
VANDEN DRIESSCHE, TH. 1966. *Exp. Cell. Res.* **42:**18–30.–
VOLM, M. 1964. *Z. vergl. Physiol.* **48:**157–180.
WASSERMANN, L. 1959. *Planta (Berlin)* **53:**647–669.–
WINFREE, A. T. 1970. *J. theor. Biol.* **28:** 327–374.

4. Autonomy of Cells and Organs; Controlling Systems

a. Independent oscillations in unicellular organisms, tissues, and organs

Unicellular organisms. The mechanism of the circadian clock is not restricted to the complex interaction of different tissues and organs within an organism. This is demonstrated very clearly by the existence of diurnal oscillations in unicellular organisms. In *Euglena* an endodiurnal rhythm of phototactic sensitivity has been carefully studied (Fig. 36; BRUCE and PITTENDRIGH 1956, 1957). *Euglena* simultaneously shows a persistent diurnal rhythm in swimming activity (BRINKMANN). The unicellular, luminescent alga *Gonyaulax* (Fig. 10) should also be mentioned again here: not only is luminescence controlled by the physiological clock (it continues in a diurnal pattern under constant conditions), but also cell division and photosynthetic capacity.

Paramecium shows rhythmical changes in sexual reactivity (SONNEBORN; EHRET; BARNETT; KARAKASHIAN). The reactive phase is limited to a certain time of day, which differs with species and varieties. Unicellular algae (*Chlamydomonas*) also have this type of circadian changes in sexual reactivity (HARTMANN).

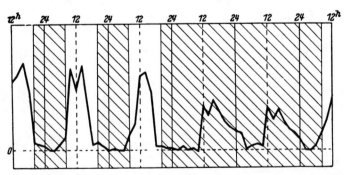

Fig. 36. Phototactic response of *Euglena gracilis* in LD of 12:12 hours followed by DD. After POHL

Autonomy of Cells and Organs; Controlling Systems 49

Isolated tissues, tissue cultures, and organs. For plants it has been known for some time that the clock does not depend on the interaction of the different parts of an individual. In isolated sections from leaves we can observe the continuation of the diurnal fluctuations in turgor pressure and growth. Even when leaf joints were halved, these fluctuations could still be identified in the isolated upper and lower sections.

This is also the case with strips isolated from petals. They grow unevenly, resulting in curvatures that coincide with the regular opening and closing movements of those flowers (BÜNNING 1942). According to investigations by WILKINS (Fig. 37), tissues isolated from leaves continue to show rhythmical changes in metabolic activity (CO_2 output) for several days. Under suitable conditions even plant tissue cultures may clearly show endodiurnal fluctuations in their turgor pressure and growth rate (Fig. 38). Much clearer are results published by WILKINS and HOLOWINSKY: leaf callus cultures of a *Bryophyllum* species possess a circadian rhythm of CO_2 metabolism. This rhythm persists in DD. Data available also for animal tissue cultures show, for example, the continuation of the rhythm in frequency of mitosis (HUPE u. GROPP; BADRAN and LLANOS). Diurnal oscillations in higher animals may proceed independently in different organs. This can be concluded from desynchronizations, i.e., from the fact that the normal phase relationships between the rhythmic functions of the various organs are lost. Isolated organs and organ parts show this independence still better. Experiments of this kind were done with sections from the intestines of hamsters. If isolated gut sections about 0.5 to 1.0 cm in

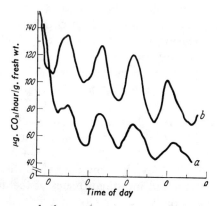

Fig. 37. CO_2 output rhythm in leaves of *Bryophyllum fedtschenkoi* in DD. (a) in leaves without the lower epidermis. (b) in pieces of mesophyll about 1 cm², from which the epidermis has been removed. After WILKINS 1959

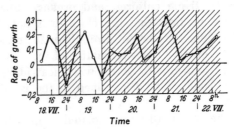

Fig. 38. Circadian changes in the growth rate of a tissue culture of *Daucus carota* (carrot). The culture was initially exposed to a regular light-dark regime, than to DD (negative "growth rates" indicate simultaneous changes in turgor pressure). After ENDERLE

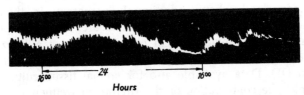

Fig. 39. *Mesocricetus auratus* (hamster). Persistence of diurnal periodicity of contractions in excised segments (7 mm long) of intestine. After BÜNNING 1958

length were placed in a suitable solution and supplied with oxygen, they remained alive for about three days if the temperature was not too high (20°C). They showed a distinct diurnal rhythm in their peristaltic activity (Fig. 39).

FURUYA and YUGARI observed a daily rhythmic change in the transport of histidine by everted sacs of the small intestine of rats, persisting for at least one day of fasting. UNGAR and HALBERG (1962, see also UNGAR) found circadian rhythms in the *in vitro* response of the mouse adrenal to adrenocorticotropic hormone (ACTH), and, again *in vitro*, rhythms in the adrenocorticotropic hypophysis activity (1963). Prolonged isolation and culture of the adrenal glands from hamsters and brown lemmings showed continuing circadian patterns in respiration, membrane flux, secretion, and in reactivity to ACTH under these conditions (ANDREWS; ANDREWS and FOLK; ANDREWS et al.).

STRUMWASSER demonstrated that the spike output of a nerve cell in the isolated parieto-visceral ganglion of the sea hare (*Aplysia californica*) fluctuates with a circadian rhythm, when the animals were in LL for one week. Brain and intestinal tissues of the arachnid *Leiobunum longipes* show a daily rhythmicity in production of the neurosecretory 5-hydroxy-tryptamine after 80-day culture (FOWLER and GOODNIGHT). The isolated eye of the seahare *Aplysia* shows a circadian rhythm of optic nerve activity when kept in DD (JACKLET; ESKIN). The circadian rhythm in the nuclear volume of isolated salivary glands from *Drosophila* larvae persists up to 10 days in a chemically defined medium (RENSING, 1969a).

From all these and other facts we may conclude that cells, tissues, or organs within multicellular organisms are able to oscillate independently. On the other hand, of course, to ensure a normal physiological functioning, the several organs, tissues, and cells have to oscillate in specific phase relationships. It is an open question whether this relationship is maintained by mutual synchronization alone or whether there are certain centers in the body with a specific control function.

b. Mutual entrainment in plants

The several organs (for example, leaves) of an individual plant can oscillate independently of each other. Plants certainly do not have a specialized organ that might be able to establish a synchronism in such cases. But a mutual synchronization and entrainment among the constituent parts exists also in plants to a certain degree, as has been shown by studying leaf movements. One leaf may influence the phase and period of another leaf (GUREVITCH; GUREVITCH and IOFFE). For example, in the soybean, *Glycine max*, one leaf with periods entrained by LD cycles to 24-hour periods can transport this information in a downward direction, i.e., entrain other leaves (showing in DD free running 26.1-hour periods) to the 24-hour period. This transportation of information is incomplete upwards (KÜBLER).

c. Controlling organs in lower animals

Annelids. Contraction and expansion of the chromatophores in the polychaete annelid *Platynereis dumerilii* show a diurnal rhythm which persists in LL or DD. Decapitation does not prevent this rhythm, showing that cerebral ganglion and eyes are not necessary (FISCHER). There is experimental evidence for assuming an autonomous capacity of circadian rhythmicity in every chromatophore (RÖSELER).

Crustaceans. In early research DEMOLL supposed that the color changes in arthropods are controlled by periodic phenomena in the nervous system. WELSH (1930, see WELSH 1938) found that the eyes of Crustaceae are important in the control by LD cycles, and he assumed that the blood is responsible for the transition of the stimulus. In 1928, KOLLER demonstrated the role of hormones. KLEINHOLZ (1937) obtained pigment migrations by injecting extracts from the eyestalks, and the sinus glands (which are included in eyestalks) were recognized as the site of secretion of the decisive melanophoric hormone. This left the possibility that the clock is located in the gland itself, resulting in a rhythmical secretion, or the gland could be controlled by a rhythmical nervous activity (WELSH 1936; see WELSH 1938). However, a rhythmical variation in the level of the responsible hormone could not be proven (KLEINHOLZ 1937; ABRAMOWITZ 1937). More recent studies with several species of crustaceans show that the eyestalks are not necessary for the known circadian rhythms in the body (BROWN et al.; BLISS; FINGERMAN and OGURO; FINGERMAN and YAMAMOTO; NAYLOR and WILLIAMS; *ROBERTS). Hormonal and nervous influences from the eyestalk may contribute to the *manifestation* and the *level* of some functions; but the fact that these functions are rhythmical must have another origin.

Insects. A controlling influence of insect periodicity by hormone release from organs in the head area is indicated by observations of KALMUS and also by some other reports published years ago. Studying the diurnal color changes in the walking-stick insect, JANDA supposed correctly that the head has an important control function. An influence of the corpora allata has been emphasized in insects (*WIGGELSWORTH; HANSTRÖM).

The data available from more recent publications are still somewhat contradictory. However, quite a number of facts may be summarized. Circadian activities of neurosecretory cells in the brain, as well as in several glands, do exist. These oscillations become evident by changes in the nuclear and nucleolar volumes and more directly in the pattern of hormone production (RENSING 1966, 1969b). In several cases a correlation between these cycles of neurosecretory activity and other circadian cycles of the insect becomes clear. For example, the circadian rhythm of running activity in the beetle *Carabus nemoralis* is paralleled by a periodical volume change in the nuclei of the corpora allata. The increase of the nuclear volume coincides with the beginning of the running activity (Figs. 40, 41). There is also a correlation between the locomotor activity and the histochemical changes in the neurosecretory cells of the pars intercerebralis and subesophageal ganglion (CYMBOROWSKI: experiments with the house cricket *Acheta domesticus*). It should be mentioned that most of these measurements refer to animals or isolated glands (RENSING 1969a) kept in diurnal LD cycles.

As with the crustaceans, with insects, too, there is no proof for theories suggesting any one of the glands to be the site of the "master clock." *HARKER assumed the subesophageal ganglia of the cockroach

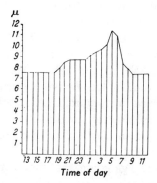

Fig. 40. Diurnal change in nuclear volume of cells of the corpora allata from the beetle *Carabus nemoralis*. After KLUG

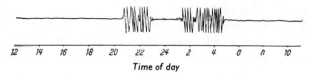

Fig. 41. Running activity of *Carabus nemoralis* under approximately normal conditions. Measurements were taken in May. After KLUG

Periplaneta americana to function in such a way. But ROBERTS, as well as BRADY (1967), were not able to confirm HARKER's results and conclusions. The activity rhythm continues in the absence of the neurosecretory cells of the subesophageal ganglion. HARKER's conclusions are mainly based on parabiosis experiments and on observing reinitiation of rhythmicity in arrhythmic individuals. But experiments such as these are subject to several sources of error (THOMAS and FINLAYSON; CYMBOROWSKI and BRADY).

Cockroaches can maintain their rhythmicity in locomotor activity after excision of the corpora cardiaca (ROBERTS) and after the almost complete elimination of the medial neurosecretory cells from the brain. NISHIITSUTSUJI-UWO and PITTENDRIGH, also working with cockroaches, conclude from their experiments that the subesophageal ganglion does not control the locomotory rhythm by a rhythmic hormone secretion. But the ganglion controls the activity *level*.

All these results concerning cockroaches agree with EIDMANN's finding for the walking-stick insect *Carausius morosus*. The activity rhythm of this animal could not be changed by removal of the corpora allata and cardiaca, lobi optici, and not even by implantation of brain or subesophageal ganglia. Disconnecting the feeler nerves or disrupting the connection between brain and frontal ganglion also failed to cause any interference. Yet the extirpation of the brain or its front part alone *did* abolish the rhythm. This was not restored by reimplantation. Separating the connectives reaching from the brain to the subesophageal ganglia also suppressed the rhythm. FINGERMAN et al. (1958) investigated the diurnal rhythm in locomotive activity and CO_2 release of the grasshopper *Romalea microptera*. Their results do not support the importance of the subesophageal ganglia either.

Similarly, the circadian rhythm of emergence from the pupal cuticle (ecdysis) apparently does not depend on such organs. Sectioning the subesophageal ganglion, the frontal ganglion, or the corpora allata and corpora cardiaca did not inhibit this process in silkworms (TRUMAN and RIDDIFORD).

Of course, inhibition of a rhythm by damaging or removing some organ in any case is insufficient to prove the location of a controlling central clock in this organ. Many organs supplying hormones may be necessary for the *manifestation* of the rhythmicity. Locomotor rhythms especially are easily disturbed (masked) or abolished by surgical operations, whereas metabolic rhythms may survive these operations.

On the other hand, when the rhythmicity continues without a specific organ, this does not prove the absence of a clock in that organ. The presence of autonomous clocks in very different organs in connection with the experiments in annelids and arthropods does not exclude the possibility that certain organs are especially important in controlling the appropriate phase relationships within the whole body.

According to the experiments by NISHIITSUTSUJI-UWO and PITTENDRIGH, the driving oscillation, controlling the locomotor rhythm of the cockroach is located in the brain, probably in the optic lobes. The optic lobes are also important in controlling the circadian stridulatory rhythm of the cricket *Teleogryllus commodus*. After severance of both optic lobes, stridulation is randomly distributed over the 24-hour period (LOHER). But a generalization can not be made from these findings. The circadian rhythmicity of stridulatory activity in species of the katydid *Ephippiger*, however, continues when the influence of the optic lobes is eliminated (DUMORTIER). (Stridulation is the sound produced by rubbing the edges of the front wings together).

*BRADY concludes that the primary circadian pacemakers for the locomotory rhythm assert their effect electrically through neural pathways. This theory finds further support from work by ROBERTS et al. (1971), again based on studies on a cockroach (see also BRADY 1971). AZARJAN and TYSHCHENKO, working with the cricket *Gryllus domesticus*, draw the same conclusion.

Apparently, the known hormonal rhythms are not primary driving oscillators, but are driven. As far as locomotor rhythms are concerned, we concur with BRADY's (1971) conclusions: "No rhythms are transferable by gland transplant, virtually all the cephalic endocrine organs can be removed without stopping the rhythm; the only operations that do stop the rhythm are those between the optic lobes and the brain, or between the brain and the thorax."

On the other hand, driven hormonal rhythms may exert a phase-shifting effect on certain overt rhythms. An interesting example for such an overt rhythm being fully dependent on rhythmic hormone supply is described by TRUMAN and RIDDIFORD. We already mentioned the lack of any influence of certain ganglia on ecdysis in silkworms. In addition, the authors were able to show the decisive role of the brain in synchronizing the rhythmic ecdysis with the LD cycles. The experiments were performed with three species of silkworms: *Hyalophora cecropia, Antheraea polyphemus,* and *A. pernyi*. Brainless moths emerged randomly throughout the day and night. Removing the brain and reimplanting it into the abdomen reestablished the periodicity. Further experiments allowed the conclusion that the LD cycles to which the brain had been exposed before implantation determined the time of ecdysis. Since the brain exerts this controlling influence even while lying loose in the abdomen, it must exert its control through hormones.

d. Controlling organs in vertebrates

General observations. The synchronous pattern and appropriate phase relationships of several diurnal functions seem to indicate a strict central control of diurnal phenomena in vertebrates, including human beings (Fig. 42). But these experiments are not convincing, since they did not show whether all these functions will also continue under constant conditions and whether the different functions keep the synchronous pattern. It is now known that even desynchronization or phase shifts may occur among the different rhythms. We shall present more details on this subject later. It should be remembered that some, but not all, diurnal functions are controlled by the endogenous diurnal rhythm. Others are completely controlled by exogenous factors; they change immediately in response to different environmental conditions. The investigations by KLEITMAN and KLEITMAN may serve as an example. The diurnal fluctuations in pulse frequency and blood pressure were strictly exogenous in humans who had not been exposed to the 24-hour rhythm of the environment. On the other hand, LEWIS and LOBBAN found in experiments under abnormal conditions that the

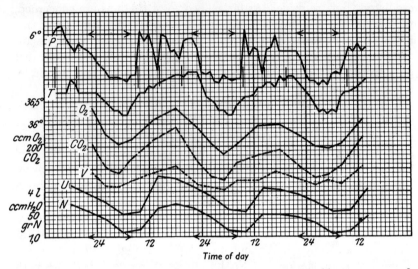

Fig. 42. The course of some diurnal functions in man. Double arrows, period of sleep in the subject; vertical lines, mealtimes; P, pulse rate; T, temperature; O_2, oxygen consumption; CO_2, CO_2 production; V, ventilation; U, urine excretion; N, nitrogen content of urine. After VÖLKER

endogenous diurnal rhythm to some extent controls the excretion of water Cl^-, and Na^+, and it entirely controls the secretion of K^+. The experiments by ASCHOFF u. WEVER mentioned earlier also indicate the existence of endogenous components (Fig. 11).

Some diurnal fluctuations in the body of vertebrates and humans are actually controlled by the circadian clock. For these, one might consider a controlling influence by brain and central nervous system. But these organs do not have as much of a central control as was originally thought. Indeed, the rhythms of different organs in higher animals are synchronized usually through the eyes and the central nervous system. But several rhythms can also proceed independently of brain and central nervous system (see Chapter 3(a)).

> **Earlier investigations.** YOUNG demonstrated the controlling influence of the pituitary gland on the movement of melanophores in *Lampetra planeri*. Projecting observations in different animals to human beings, JORES postulated a decisive role of the pituitary. The hormones of the pituitary in the gland and in the blood fluctuate diurnally. Since these fluctuations are controlled by the changes of light and darkness, they may well be the basis of some physiological phenomena that are dependent on the LD cycle (see also °MENZEL). JORES assumed that the increase in body temperature, pulse rate, blood pressure, and oxygen-consumption are paralleled by an increase of adrenaline in the blood. And the opposite phase during the night

is marked by an increase in pituitary hormone. Indications of an increased adrenal activity during the day and an increased pituitary activity during the night are given in reports by several writers (EULER u. HOLMQUIST).

Adrenal gland. HALBERG thoroughly investigated the influence of periodical activity changes in the adrenal gland. The adrenal cycle controls various diurnal functions in the body of vertebrates, including humans. The number of eosinophils circulating in the blood reflects the course of this adrenal cycle. The hormones of the adrenal gland tend to diminish the eosinophils, thereby controlling their number. An extirpation, or pathological loss, of the adrenal gland function abolishes the eosinophil rhythm (Fig. 43). From this, HALBERG concludes a 24-hour rhythm in hormone secretion. An extirpation of the adrenal gland also abolished the glycogen rhythm in the liver (AGREN et al.).

There is a diurnal rhythm in the excretion of adrenal cortex hormone. PINCUS (1943) found evidence of a 24-hour rhythm in the excretion of 17-ketosteroids, which are characteristic of these hormones. Also, the expected antagonism between eosinophils and ketosteroids appears (Fig. 44). On the basis of the investigations by HALBERG and coworkers, we may regard the ketosteroid excretion as a function controlled by the endogenous diurnal rhythm. Mice in DD maintain the adrenal cycle for several weeks.

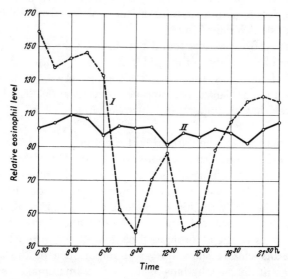

Fig. 43. Obliterations of eosinophil rhythm in man in verified adrenocortical insufficiency. *I*, unlimited activity (adrenal sufficiency); *II*, limited activity (adrenal insufficiency). After HALBERG 1953

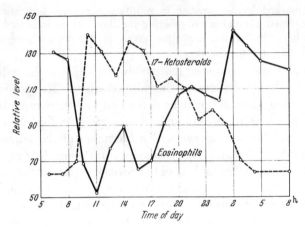

Fig. 44. Circadian rhythm in circulating eosinophils and in excretion of 17 ketosteroids in man. After HALBERG 1953

The central nervous system is not required for the adrenal cycle. This cycle also appears in brainless animals (GALICICH). Furthermore, UNGAR and HALBERG found that the removal time of the adrenal gland influences the *in vitro* production of corticosterone by the adrenal cortex after stimulating it with adrenocorticotropic hormone (ACTH). The rhythm, therefore, is not caused simply by the central nervous system and by a periodical ACTH production in the pituitary gland. The continuation of the rhythms in the isolate adrenal was already mentioned (see Chapter 3(a)).

HALBERG and coworkers found that in mice the adrenal cycle controls the mitotic activity in the epidermis as well as a number of other physiological functions; for example, running activity is paralleled by the initial increase in corticosterone concentration. Evidently this adrenal cycle has an important controlling influence on diurnal processes.

The periodicity of the adrenal cortex activity can be identified not only by its effects, but also in a more direct way. The gland itself shows a periodical change in hormone concentration and a diurnal periodicity in mitosis (Fig. 45). The most active hormone secretion of the adrenal cortex is paralleled by the highest mitotic rate. GLICK *et al.* observed a periodicity in enzyme activity, and RINNE and KYTÖMÄKI noticed diurnal changes in the ascorbic acid content of the adrenal gland.

The results clearly show the great importance of the adrenal in rather different diurnal rhythms of the body in mammals. Yet, it is not certain whether the cycles inside the gland are pacemakers to all the connected rhythms outside the gland, or whether they are predomi-

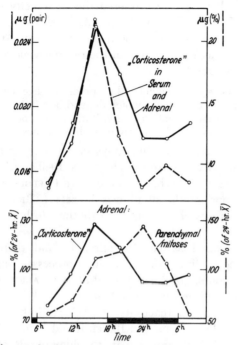

Fig. 45. Circadian rhythm of corticosterone in mouse blood and adrenal in relation to mitosis in glandular parenchyma. After HALBERG et al. 1959

nantly only important members in the system of mutual synchronization (see the respective remarks for arthropods). A rhythm like running activity, though clearly related to the adrenal cycle, persists after the removal of the adrenal (*RICHTER, 1965). The same holds with respect to the mitotic cycle. Adrenalectomy did not abolish this cycle in the corneal epithelium of the rat (SCHEVING and PAULY).

In any case, the adrenal cycle again is synchronized by external cues, especially by LD cycles. We do not know as yet very much about the complicated processes by which the eye or other light-absorbing organs and the central nervous system can control the activity of the gland (see also Chapter 6).

Thyroid gland. The thyroid or small parts of it favor the rhythm of running activity; but removing the thyroid does not erase the rhythm (KALMUS 1940; RICHTER 1933, *1965). Furthermore, diurnal rhythms in body temperature are only damped after thyroidectomy (POPOVIC et PETROVIC).

Pituitary gland. Several authors have repeatedly discussed the participation of the pituitary (hypophysis) in controlling periodical functions. Some have even considered the periodicity of the adrenal activity as being controlled by the pituitary. However, in the "circadian organization" of the body, both the rhythm of ACTH secretion by the hypophysis and the rhythm of the responsiveness of the adrenal to ACTH are important.

> It is nearly self-evident that processes like color changes, in several species of animals, which depend on pituitary hormones, will be stopped after hypophysectomy. But this does not necessarily mean that normally a periodic supply of such hormones is necessary.

Rhythms continuing after hypophysectomy are, for example, those of locomotor activity in *Axolotl* (KALMUS 1940), in rats and mice (MÜLLER u. GIERSBERG), in body temperature (FERGUSON et al.), blood corticosteroids, and epithelial mitosis in rats (*RICHTER; *ROBERTS; see also PAULY's research on leukocyte rhythms in hypophysectomized rats).

On the other hand, as with the adrenal gland, there are certainly rhythmical processes in the pituitary. Such processes have been shown with the ACTH content of the mouse pituitary (GALICICH et al.), which has also been established *in vitro* (HALBERG et al. 1965). Thus, there is not only a rhythmic reactivity to ACTH, but also a rhythmic production of ACTH. CLARK and BAKER found a circadian periodicity in the concentration of prolactin in the hypophysis of rats, with a 325% increase occurring early in the afternoon and its lowest point a few hours before midnight. But the rat's body also has a rhythmic responsiveness to prolactin (MEIER et al.). The rat hypophysis also shows a circadian periodicity in thyroid stimulating hormone (BAKKE and LAWRENCE).

Finally, there is a good deal of information about rhythmic activity of the pituitary in connection with ovulation cycles and related phenomena. There is much evidence for diurnal rhythm in secretion of the several gonadotropic hormones, and there is a typical phase relationship between the release of the several hormones and other events during the ovulation cycle. EVERETT and SAWYER recognized the diurnal rhythmicity in secretion of luteinizing hormones by the hypophysis in rats. In birds, there is a similar circadian rhythm in the release of gonadotropic hormones (FRAPS 1954, *1959, *1961; see also *WILSON). Periodic hormone secretion may be demonstrated with just one example. Injection of gonadotrophin induces puberty in rats. Yet ovulation does not occur at a fixed time after the injection. The hypophysis secretes hormones only during the night (from 2 to 4 A.M. under the special experimental conditions). Applying barbituric acid to the nervous system blocked ovulation temporarily, delaying it by 24 hours. A second dose of barbituric acid on the next day delayed ovulation by two days compared to the controls. Ovulation was prevented entirely by removing the hypophysis before 2 A.M., but it could not be prevented if it was removed after 4 A.M. Obviously, the ovulation hormone is secreted by the hypophysis according to a diurnal rhythm. (STRAUSS and MEYER; see also H. E. BROWN.) On the other hand, there is a circadian responsiveness to injected gonadotrophin in hypophysectomized mice (BINDON and LAMOND).

> It is beyond the intentions of this book to go into further details of the rather complicated cooperation between pituitary, follicles, nervous system, and external cues like LD cycles, which are involved in ovulation cycles. Only the rhythmic functioning of the pituitary should be mentioned here (see also KENDALL and ALLEN, and the reviews in *HAGUE, Ed.).

Hypothalamus. The pituitary cycles apparently do not run autonomously (as was assumed by earlier hypotheses, see HILL and PARKES). They are predominantly under control of the central nervous system. The pituitary ACTH rhythm is strongly influenced by brain ablations, like removal of the hypothalamus (GALICICH et al.). Other evidence for a hypothalamic rhythmicity is also known, and more generally, the participation of neural components in diurnal cycles like those of ovulation seems to be necessary (with respect to birds see FRAPS 1965, OPEL; with respect to mammals see ROBERTSON and RAKHA; WURTMAN et al. 1964; *RETIENE).

STEPHAN and ZUCKER found an inhibition of the circadian rhythms of drinking and locomotor activity of rats after bilateral lesions in the region of the hypothalamus' suprachiasmatic nuclei. Interruption beyond the chiasm of the primary and inferior accessory optic systems did not disturb the entrainment of the drinking rhythm by LD cycles. These authors suggest that the suprachiasmatic region of the rat brain may contain pacemakers for circadian and other behavioral rhythms.

Many of the cycles which are influenced by pituitary and hypothalamus can be controlled by LD cycles. Again, it cannot be our task here to discuss how the external diurnal cycles synchronize the cycles in these organs. It may just be mentioned that in mammals not only the eyes, but also the central nervous system (hypothalamus) itself can function as a decisive light absorbing system (LISK and KANN-WISCHER). We will have to mention these light effects once more in connection with photoperiodic control phenomena (Chapter 13; reviews in *HAGUE, *BENOIT et ASSENMACHER.

Pineal gland. The pineal gland of rats shows circadian rhythms in serotonin with highest levels at noon and lowest at midnight (QUAY, 1963 and 1966; SNYDER et al.). This, too, is not an autonomous rhythm of the gland itself. It is not affected by hypophysectomy, thyroidectomy, and adrenalectomy, but it is under control of the central nervous system. Through this system, LD cycles influence the rhythmic activity of the pineal. The serotonin cycle of rats can persist in DD or in blinded animals. The diurnal cycle in melatonin, however, is apparently truly exogenous, i.e., entirely dependent on the environmental lighting. *WURTMAN and AXELROD assume that the pineal clock may be especially important in the timing of the estrus and menstrual cycles.

In the house sparrow (*Passer domesticus*) and in the white-throated sparrow (*Zonotrichia albicollis*), the pineal organ is essential for the persistence of the locomotor and other circadian rhythms (GASTON and MENAKER; GASTON; BINKLEY et al.; MCMILLAN).

Yet, the pineal does not shelter a "master clock," the running of which is necessary for all the other rhythms of the body. Even if the pineal is a pacemaker for the locomotor rhythm in birds, this is not a general rule. Pinealectomy does not result in a loss of the activity rhythm in reptiles (UNDERWOOD and MENAKER) and in mammals (*RICHTER 1967; QUAY 1968).

The main role of the pineal, as far as the clock is concerned, appears to be the involvement in entrainment by LD cycles, and in internal synchronization. According to ADLER (1971), there are good reasons to assume a photoreceptor function of the pineal end organ in the green frog, *Rana clamitans*.

Circadian rhythms in the blood. The synchronization of some (but certainly not all) of the functions in the body of vertebrates is possible through the mediation by the central nervous system and by transport of hormones in the blood. Therefore, in addition to the blood rhythm we mentioned already, there is a great number of other circadian rhythms in physicochemical parameters of the blood, as has been reviewed together with additional information by *SEAMAN et al. (See also FEIGIN et al., BAHORSKY and BERNARDIS.) Many of these cycles might be exogenous—for example, depending on food supply. But others show the features of circadian rhythms. Thus, WURTMAN et al. (1967) found a rhythm in tyrosine concentration in human plasma, persisting when the subjects were living for two weeks on a diet very low in protein.

Controlling factors mediated by the blood can also be detected by other phenomena, such as the behavior of certain blood parasites. The developmental cycle of the bird parasite *Plasmodium cathemerium* (malaria) shifts its phases when the birds are exposed to an inverse LD cycle (BOYD). The same applies to the diurnal oocystis production by *Isospora* in the sparrow (BOUGHTON et al.). Evidently the periodical appearance of microfilariae larvae in human blood is also controlled by the blood. It reverses if the living habits of a person are reversed

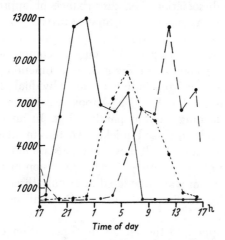

Fig. 46. Inversion of the number of filaria (*Microfilaria nocturna*) in blood by shift in sleeping habit due to variation in light period. ———, sleep from 7 P.M. to 7 A.M.; - - - -, 4th day of sleep from 7 A.M. to 7 P.M.; — — —, 11th day of sleep from 7 A.M. to 7 P.M.; ordinate, number of filaria per ml blood. After YORKE and BLALOCK 1917, from *ASCHOFF 1955

(Fig. 46). For further details on 24-hour rhythms in blood parasites, see ENGEL et al., HAWKING, *HAWKING 1962, 1970.

The circadian organization. The facts and hypotheses stated above show the difficulties of analyzing processes in the bodies of higher animals. But generally there is no doubt that the individual organs and tissues in higher animals, just as those in plants and lower animals, are still capable of endodiurnal oscillations. On the other hand, some organs exert a strong controlling effect and are responsible for the synchronization of processes in different parts of the body, as well as for the mediation of external cues (LD cycles).

"Synchronization" does not necessarily mean the time coincidence of maximum activity in different organs and tissues. It only means that a certain phase relation necessary for the normal functioning of the body is maintained. Actually, the several rhythms in the body of a mammal show similar periods, but their peaks or troughs do not all occur at the same time (*HALBERG; *HALBERG et al.).

There is a great multiplicity of circadian oscillations in single cells, tissues, and organs of higher animals. The examples mentioned in this chapter are no more than a very small fraction of what is known. For more detailed information, see the reviews quoted in the references to this chapter and to Chapter 1.

Dissociations, splitting. Other evidence for a relative independence of the several organs and physiological functions inside the body are observations on dissociation—i.e., disturbance of appropriate normal phase relationships. As one example, SHARP's investigations with humans may be mentioned. After several days of LL, persons were exposed to a LD cycle that was shifted by 12 hours in comparison to previous conditions. The inversion of different diurnal functions took a certain length of time varying from individual to individual. The reversal of excretion rhythms (water, K^+, Na^+) took two to nine days, and the reversal of the sleeping rhythm paralleled it. In an easily adapting person, all these cycles reversed quickly, whereas in others these cycles reversed slowly. Apparently all these functions are more or less subject to a common control in the body. Yet the inversion of the lymphocyte and eosinophil rhythm was completed after only three days or earlier. Additional facts concerning such dissociations will be discussed later on (Chapter 14).

What may happen to the circadian organization of the body in case of missing control by external cues is clear from rhythm dissociations occurring in blinded animals. The role of the normal pathway of LD cycle effects via the eyes, hypothalamus, and pituitary in higher animals may also be demonstrated by experiments with hypophysectomized mice (FERGUSON et al.). The rhythm of body temperature continues, but it is no longer synchronized with the LD cycles.

Not only different physiological functions might be separated in time from one another. Under certain conditions, even the recording of one type of rhythmicity may show a splitting into two rhythms with different frequencies. HOFFMANN (1969, 1971) observed such a splitting when recording the locomotor activity of *Tupaia glis* after lowering the light intensity to about one lux, (Fig. 133). Raising the light intensity resulted in fusing the two components again (see also PITTENDRIGH, and H. POHL for similar observations).

A splitting such as this may even be connected with differences in the period length of the two parts (HOFFMANN 1971). This fact clearly shows that there is more than just *one* control center for locomotor activity.

> When discussing pathological phenomena (Chapter 14), we will once more have to describe consequences of missing synchronization because of lack of effective information from the outside or because of disturbed mutual synchronization inside the body.

References to Chapter 4
(Autonomy of Cells and Organs: Controlling Systems)

a. Reviews

*Aschoff, J. 1955. Exogene und endogene Komponente det 24-Stunden-Periodik bei Tier und Mensch. *Naturwiss.* **42**:569–575.

*Benoit, J., et I. Assenmacher, eds. 1970. *La Photorégulation de la Reproduction chez les Oiseaux et les Mammifères.* Paris: Centr. Nat. Rech. Scientif.—

*Brady, J. 1969. How are insect circadian rhythms controlled? *Nature* **223**: 781–784.

*Fraps, R. M. 1959. Photoperiodism in the female domestic fowl. In *Photoperiodism and Related Phenomena in Plants and Animals.* R. B. Withrow, ed. Washington: Am. Ass. Adv. Sci. Pp. 767–785;—

———. 1961. Ovulation in the domestic fowl. In *Control of Ovulation.* Claude A. Villee, ed. Oxford: Pergamon. Pp. 133–162.

*Hague, E. B., ed. 1964. Photo-Neuro-Endocrine Effects in Circadian Systems, with Particular Reference to the Eye. *Ann. N. Y. Acad. Sci.* **117**: 1–645.—

*Halberg, F. 1959. Physiological 24-hour-periodicity; general and procedural considerations with reference to the adrenal cycle. *Z. Vitamin-Hormon-Fermentforsch.* **10**:225–296;—

———. 1960. Temporal organization of physiologic function. *Cold Spring Harbor Symp. quant. Biol.* **25**:289–310;—

———. 1963. Periodicity analysis a potential tool for biometeorologists. *Int. J. Biometeor.* **7**:167–191.—

*Halberg, F., E. Halberg, C. P. Barnum, and J. J. Bittner. 1959. Physiologic 24-hour periodicity in human beings and mice, the lighting regimen and daily routine. In *Photoperiodism and Related Phenomena in Plants and Animals.* R. B. Withrow, ed. Washington: Am. Ass. Adv. Sci. Pp. 803–878.—

*Harker, J. 1960. Endocrine and nervous factors in insect circadian rhythms. *Cold Spring Harbor Symp. quant. Biol.* **25**:279–287.—

*Hawking, F. 1962. *Microfilaria* infestation as an instance of periodic phenomena seen in host-parasite relationships. *Ann. N. Y. Acad. Sci.* **98**: 940–953;—

———. 1970. The clock of the malaria parasite. *Scientific American* **222**: 123–131.

*Menzel, W. 1952. Über den heutigen Stand der Rhythmuslehre in bezug auf die Medizin. *Z. Altersforsch.* **6**:26–121.

*Retiene, K. 1970. Control of circadian periodicities in pituitary function. In *The Hypothalamus.* L. Martini, M. Motta, and F. Fraschini, eds. New York–London: Academic Press. Pp. 551–568.—

°RICHTER, C. P. 1965. Biological Clocks in Medicine and Psychiatry. Springfield, Ill.: Charles C Thomas;—
———. 1967. Sleep and activity: their relation to the 24-hour clock. In *Sleep and Altered States of Consciousness*. Ass. Res. Nerve and Mental Disease XLV. Baltimore: Williams & Wilkins. Pp. 8–29.—
°ROBERTS, S. K. 1965. Significance of endocrines and central nervous system in circadian rhythms. In *Circadian Clocks*. J. ASCHOFF, ed. Amsterdam: North Holland Publishing Company. Pp. 198–213.
°SEAMAN, G. V. S., R. ENGEL, R. L. SWANK, and W. HISSEN. 1965. Circadian periodicity in some physicochemical parameters of circulating blood. *Nature* 207:833–835.
°TRUMAN, J. W. 1972. Circadian rhythms and physiology with special reference to neuroendocrine processes in insects. In *Circadian Rhythmicity*. J. F. BIERHUIZEN et al., eds. Proc. Int. Symp. on Circad. Rhythmicity Wageningen. Centre for Agricult. Publ. a. Documentation. Pp. 111–135.
°WILSON, W. O. 1964. Photocontrol of oviposition in gallinaceous birds. *Ann. N. Y. Acad. Sci.* 117:194–203.—
°WURTMAN, R., and J. AXELROD. 1965. The pineal gland. *Scientif. Amer.* 213:50–60.—
°WIGGLESWORTH, V. B. 1950. *The Principles of Insect Physiology*. London: Methuen & Co.

b. Other references

ADLER, K. 1971. In *Biochronometry*. M. MENAKER, ed. Pp. 342–350. Washington, D.C.: Nat. Acad. Sci.—
ABRAMOWITX, A. A. 1937. *Biol. Bull.* 72:344–365.—
AGREN, G., O. WILANDER, and E. JORPES. 1931. *Biochem. J.* 25:777–785.—
ANDREWS, R. V. 1968. *Comp. Bioch. Physiol.* 26:179–193.—
ANDREWS, R. V., and G. E. FOLK. 1964. *Comp. Bioch. Physiol.* 11:393–409.—
ANDREWS, R. V., L. C. KEIL, and N. N. KEIL. 1968. *Acta Endocrinol.* 59:36–40.—
AZARJAN, A. G., and V. G. TYSCHENKO. 1970. *Rev. d'Entomol. de l'URSS* 49:72–82.—
BADRAN, A. M. F., and J. M. E. LLANOS. 1965. *J. Nat. Canc. Inst.* 35:285–290.—
BAHORSKY, M. S., and L. L. BERNARDIS. 1967. *Experientia* 23:634–635.—
BAKKE, J. L., and N. LAWRENCE. 1965. *Metav. Clin. Exp.* 14:841–843.—
BARNETT, A. 1966. *J. cell. Physiol.* 67:239–270.—
BINDON, B. M., and D. R. LAMOND. 1966. *J. reproduct. Fert.* 12:249–261.—
BINKLEY, S., E. KLUTH, and M. MENAKER. 1971. *Science* 174:311–314;—
———. 1972. *J. comp. Physiol.* 77:163–169.
BLISS, D. E. 1960. *Science* 132:145–147.—
BOUGHTON, D. C., F. O. ATCHLEY, and L. C. ESKRIDGE. 1936. *J. exp. Zool.* 70:55–74.—

BOYD, G. H. 1929. *J. exp. Zool.* 54:111–126.–
BRADY, J. 1967. *J. exp. Biol.* 47:153–163, 165–178;–
———. 1968. *J. exp. Biol.* 49:39–47;–
———. 1971. In *Biochronometry*. M. MENAKER, ed. Pp. 517–526. Washington, D.C.: Nat. Acad. Sci.–
BRINKMANN, K. 1966. *Planta (Berl.)* 70:344–389.–
BROWN, F. A., M. F. BENNETT, and H. M. WEBB. 1954. *J. cell. comp. Physiol.* 44:477–505.–
BROWN, H. E. 1962. *Ann. N. Y. Acad. Sci.* 98:995–1006.–
BRUCE, V. G., and C. S. PITTENDRIGH. 1956. *Proc. Nat. Acad. Sci. USA* 42:676–682;–
———. 1957. *Amer. Naturalist* 91:179–195.–
BÜNNING, E. 1942. *Z. Bot.* 37:433–486;–
———. 1958. *Naturwiss.* 45:68.
CLARK, R. H., and B. L. BAKER. 1964. *Science* 143:375–376.–
CYMBOROWSKI, B. 1970. *Zool. pol.* 20:127–149.–
CYMBOROWSKI, B., and J. BRADY. 1972. *Nature, New Biology* 236:221–222.
DEMOLL, R. 1911. *Zool. Jahrb. Physiol.* 30:159–180.–
DUMORTIER, B. 1972. *J. comp. Physiol.* 77:80–112.
EHRET, CH. 1959. *Fed. Proc.* 18:1232–1240;–
———. 1960. *Cold Spring Harbor Symp. quant. Biol.* 25:149–158.–
EIDMANN, H. 1956. *Z. vergl. Physiol.* 28:370–390.–
ENDERLE, W. 1951. *Planta (Berl.)* 39:530–588.–
ENGEL, R., F. HALBERG, W. L. DASSANAYAKE, and J. DE SILVA. 1962. *Amer. J. Trop. Med. a. Hygiene* 11:653–663.–
ESKIN, A. 1971. *Z. vergl. Physiol.* 74:353–371.–
EULER, U. B. v., u. A. G. HOLMQUIST. 1934. *Pflügers Arch. ges. Physiol.* 234:210–224.–
EVERETT, J. W., and CH. C. SAWYER. 1950. *Endocrinology* 47:198–218.
FEIGIN, R. D., H. G. DANGERFIELD, and W. R. BEISEL. 1969. *Nature* 221:94–95.–
FERGUSON, D. D., M. B. VISSCHER, F. HALBERG, and L. M. LEVY. 1957. *Amer. J. Physiol.* 190:235–238.–
FINGERMAN, M., A. D. LAGO, and M. F. LOWE. 1958. *Amer. Midland Naturalist* 59:58–66.–
FINGERMAN, M., and CH. OGURO. 1963. *Biol. Bull.* 124:24–30.–
FINGERMAN, M., and Y. YAMAMOTO. 1964. *Am. Zool.* 4:334 (quoted according to ROBERTS).–
FISCHER, A. 1965. *Z. Zellforsch.* 65:290–312.–
FOWLER, D. J., and C. J. GOODNIGHT. 1966. *Science* 152:1078–1080.–
FRAPS, R. M. 1954. *Proc. Nat. Acad. Sci. USA* 40:348–356;–
———. 1965. *Endocrinology* 77:5–18.–
FURUYA, S., and Y. YUGARI. 1971. *Bioch. Bioph. Acta* 241:245–248.
GALICICH, J. H. 1961. 39th Ross Conference on Pediatric Research. Columbus, Ohio. Pp. 44–55.–

GALICICH, J. H., F. HALBERG, L. A. FRENCH, and F. UNGAR. 1965. *Endocrinology* 76:895–901.–
GASTON, S. 1971. In *Biochronometry*. M. MENAKER, ed. Pp. 541–548. Washington, D.C.: Nat. Acad. Sci.–
GASTON, S., and M. MENAKER. 1968. *Science* 160:1125–1127.–
GLICK, D., R. B. FERGUSON, L. J. GREENBERG, and F. HALBERG. 1961. *Amer. J. Physiol.* 200:811–814.–
GUREVITCH, B. KH. 1967. *Dokl. Akad. Nauck S. S. S. R.* 173:1459–1462.–
GUREVITCH, B. KH., and A. A. IOFFE. 1970. *Bot. Journ.* (*Russ.*) 55:77–81.
HALBERG, F. 1953. *Lancet* 20–32.–
HALBERG, F., G. FRANCK, R. HARNER, J. MATTHEWS, H. AAKER, H. GRAVEN, and J. MELBY. 1961. *Experientia* (*Basel*) 17:282–284.–
HALBERG, F., J. H. GALICICH, F. UNGAR, and L. A. FRENCH. 1965. *Proc. Soc. Exp. Biol. a. Med.* 118:414–419.–
HALBERG, F., E. HALBERG, C. P. BARNUM, and J. J. BITTNER. 1959. In *Photoperiodism and Related Phenomena in Plants and Animals*. R. B. WITHROW, ed. Pp. 803–878. Washington, D.C.: Amer. Ass. Adv. Sci.–
HANSTRÖM, B. 1937. *Sv. Vetensk. Akad. Handl. III*, 16:3.–
HARTMANN, K. M. 1962. Diss. Tübingen.–
HAWKING, F. 1967. *Proc. roy. Soc. B.* 169:59–76.–
HILL, M., and A. S. PARKES. 1930. *Proc. roy. Soc. B.* 113:537–540.–
HOFFMANN, K. 1969. *Zool. Anz. Suppl.* 33:171–177;–
———. 1971. In *Biochronometry*. M. MENAKER, ed. Pp. 134–151. Washington, D.C.: Nat. Acad. Sci.–
HUPE, K., u. A. GROPP. 1957. *Z. Zellforsch.* 46:67–70.
JACKLETT, J. W. 1969. *Science* 164:562–563;–
———. 1971. In *Biochronometry*. M. MENAKER, ed. Pp. 351-362. Washington, D.C.: Nat. Acad. Sci.–
JANDA, V. 1934. *Vestu Kràl. Cos. Spol. nauk. Praha II*, tr. 44.–
JORES, A. 1938. *Dtsch. med. Wschr.* Nr. 21 u. 28.
KALMUS, H. 1935. *Biol. Gen.* 11:93–114;–
———. 1940. *Nature* 145:42.–
KARAKASHIAN, M. W. 1968. *J. cell. Physiol.* 71:197–209.–
KENDALL, J. W., and C. F. ALLEN. 1967. *Nature* 215:876–877.–
KLEINHOLZ, L. H. 1937. *Biol. Bull.* 72:24–36.–
KLEITMAN, N., and E. KLEITMAN. 1953. *J. appl. Physiol.* 6:283–291.–
KLUG, H. 1958/59. *Wiss. Z. Humboldt-Univ. Berlin* 8:405–434.–
KOLLER, G. 1928. *Z. vgl. Physiol.* 8:601–612.–
KÜBLER, F. 1969. *Z. Pflanzenphysiol.* 61:310–313.
LEWIS, P. R., and M. C. LOBBAN. 1956. *J. Physiol.* 133:670–680;–
———. 1957. *J. exp. Physiol.* 42:356–371.–
LISK, R. D., and L. R. KANNWISCHER. 1964. *Science* 146:272–273.
LOHER, W. 1972. *J. comp. Physiol.* 79:173–190.
MCMILLAN, J. P. 1972. *J. comp. Physiol.* 79:105–112;–
MEIER, A. H., K. B. DAVIS, and R. LEE. 1967. *Gen. a. comp. Endocrinol.* 8:110–114.–

MÜLLER, M., u. H. GIERSBERG. 1957. Z. vergl. Physiol. 40:454–472.
NAYLOR, E., and B. G. WILLIAMS. 1968. J. exp. Biol. 49:107–116.–
NISKIITSUTSUJO-UWO, J., and C. S. PITTENDRIGH. 1968. Z. vergl. Physiol. 58: 1–13; 14–46.
OPEL, H. 1964. Endocrinology 74:193–200.
PAULY, J. E. 1965. Anat. Rec. 153:349–360.–
PINCUS, G. 1943. Journ. clin. Endocrinol. 3:195–199.–
PITTENDRIGH, C. S. 1960. Cold Spring Harbor Symp. quant. Biol. 25:159–184.–
POHL, H. 1972. J. comp. Physiol. 78:60–74.–
POHL, R. 1948. Z. Naturforsch. 3b:367–374.–
POPOVIC, P., et V. PETROVIC. 1956. Compt. Rend. Soc. Biol. Paris 150:1249.
QUAY, W. B. 1963. Gen. a. comp. Endocrinol. 3:473–479;–
———. 1966. Gen. a. comp. Endocrinol. 6:371–377;–
———. 1968. Physiol. Behav. 3:109–118.
RENSING, L. 1964. Science 144:1586–1587;–
———. 1966. Z. Zellforsch. 74:539–558;–
———. 1969a. J. Insect Physiol. 15:2285–2303;–
———. 1969b. Nachr. Akad. Wiss. Göttingen II. Jahrg. 1969:57–70.–
RENSING, L., B. THACH, and V. B. BRUCE. 1965. Experientia 21:103.
RICHTER, C. P. 1933. Endocrinology 17:73–87.–
RINNE, U. K., and O. KYTÖMÄKI. 1961. Experientia (Basel) 17:513.–
ROBERTS, S. K. 1959. Ph.D. Thesis, Princeton;–
———. 1966. J. Cell. Physiol. 67:473–486.–
ROBERTS, S. K., S. D. SKOPIK, and R. J. DRISKILL. 1971. In Biochronometry. M. MENAKER, ed. Pp. 505–516. Washington, D.C.: Nat. Acad. Sci.–
ROBERTSON, H. A., and A. M. RAKHA. 1965. J. Endocrin. 32:383–386.–
RÖSELER, I. 1970. Z. vergl. Physiol. 70:144–174.
SCHEVING, L. E., and J. E. PAULY. 1967. In The Cellular Aspects of Biorhythms. H. V. MAYERSBACH, ed. Berlin, Heidelberg. New York: Springer. Pp. 167–174.–
SHARP, G. W. G. 1962. Atti VII Conf. Soc. Ritmi Biol. Siena. Panminerva Medica Torino. Pp. 133–138.–
SNYDER, S. H., and J. AXELROD. 1965. Science 149:542–544.–
SNYDER, S. H., M. ZWEIG, J. AXELROD, and J. E. FISCHER. 1965. Proc. Nat. Acad. Sci. USA 53:301–305.–
SONNEBORN, R. M. 1938. Proc. Amer. Phil. Soc. 79:411–434.–
STEPHAN, F., and I. ZUCKER. 1972. Proc. Nat. Acad. Sci. USA 69:1583–1586.–
STRAUSS, W. F., and R. K. MEYER. 1962. Science 137:860–861.–
STRUMWASSER, F. 1965. In Circadian Clocks. J. ASCHOFF, ed. Amsterdam: North Holland Publishing Company. Pp. 44–62.
THOMAS, R., and L. H. FINLAYSON. 1970. Nature 228:577–578.–
TRUMAN, J. W., and L. M. RIDDIFORD. 1970. Science 167:1624–1626.
UNDERWOOD, H., and M. MENAKER. 1970. Science 170:190–193.–
UNGAR, F. 1964. Ann. New York Acad. Sci. 117:374–385.–
UNGAR, F., and F. HALBERG. 1962. Science 137:1058–1060.–

UNGAR, F., and F. HALBERG. 1963. *Experientia* **19**:158–160.
VÖLKER, H. 1927. *Pflügers Arch. ges. Physiol.* **215**:43–77.
WELSH, J. H. 1938. *Quart. Rev. Biol.* **13**:123–139.–
WILKINS, M. B. 1959. *J. exp. Bot.* **10**:377–390;–
———. 1960. *Cold Spring Harbor Symp. quant. Biol.* **25**:115–129.
WILKINS, M. B., and A. W. HOLOWINSKY. 1965. *Plant Physiol.* **40**:907–909.–
WURTMAN, R. J., J. AXELROD, E. W. CHU, and J. E. FISCHER. 1964. *Endocrinology* **75**:266–272.–
WURTMAN, R. J., CH. CHOU, and CH. M. ROSE. 1967. *Science* **158**:660–662.
YOUNG, J. Z. 1935. *J. exp. Biol.* **12**:254–270.

5. Temperature Effects

a. Temperature and length of period

Earlier experiments. The first impetus to examine the influence of temperature on the period was provided by the assumption that the endogenous diurnal rhythm was based on alternating chemical processes. One expected a shortening of the periods at high temperatures and a lengthening at low temperatures.

While the first experiments of this kind had the expected results, it was noticed that the temperature coefficients were unusually low. In studying the circadian leaf movements of beans (*Phaseolus coccineus*) a Q_{10} value of only 1.2 was found (BÜNNING 1931). The first experiments in animals seemed to indicate a similar temperature influence. KALMUS (1935) found a delay of several hours in the emergence periodicity of *Drosophila* flies when the temperature was lowered by 10° and an acceleration by raising the temperature (BÜNNING 1935 had similar results).

But these earlier experiments were subject to different sources of error. One problem was that the methods available did not permit completely satisfactory results. Another one was that the influence of temperature was often investigated only for one to three periods, again, because of difficulties in methods. This procedure leads to error for two reasons. First, temperature can, in fact, influence a process dependent upon the clock—for example, a change in growth, a change in turgor pressure, or the emergence of insects from their pupae, or the running activity. These direct influences on the controlled process interfere with possible influences on the clock itself. Second, this procedure can also lead to error because of the occurrence of "transients"; a transitory reaction is something entirely different from the steady state reached after only two or three cycles.

Low temperature dependence. For these reasons it became necessary to examine the periodicity during a longer-lasting influence of high or low temperature. Those experiments yielded the surprising result that the length of the periods is virtually independent of temperature.

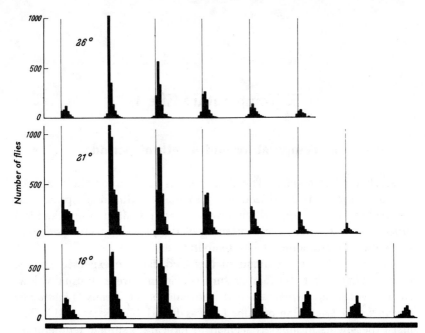

Fig. 47. Rhythm in *Drosophila* eclosion. DD at 26°, 21°, and 16°C. Broad black lines indicate darkness. Vertical guide lines are 24 hours apart. Slight influence of low temperature in lengthening the period. After PITTENDRIGH and BRUCE

This was found first in the time sense of bees. It gave rise to the idea of a compensating process working in those organisms, allowing the clock to be independent of temperature for a wide range of temperatures and to work reliably independent of normal temperature fluctuations (WAHL, see also KALMUS 1935). Temperatures between 16° and 32° had no influence. The emergence rhythm in *Drosophila* was not affected significantly by temperatures between 16° and 26° (Fig. 47).

Recently, a number of other examples showing independence of temperature or only a slight influence of temperature on the period in plants and animals have been reported. Some examples from different organisms illustrate this in Tables 1 to 4.

*SWEENEY and HASTINGS give a comprehensive list of data from many organisms, with examples ranging from unicellular plants to mammals. According to their summary there are only a few cases with a Q_{10} value higher than 1.1 or 1.2 and quite often the Q_{10} values range between 1 and 1.1 (see also *WILKINS).

Table 1. *Periplaneta americana,*
Running Activity
(BÜNNING 1958a)

Temperature	Length of Period
°C	hr
18	24 – 25
19–20	24.4 ± 0.1
22–23	24.5 ± 0.1
27–28	25.0 ± 0.3
29	25.8 ± 0.7
31	24 – 27

Table 2. *Gonyaulax polyedra,*
Rhythm of Luminescence
(HASTINGS and SWEENEY)

Temperature	Length of Period
°C	hr
15.9	22.5
19	23.0
22	25.3
26.6	26.8
32	25.5

Table 3. *Phaseolus coccineus,*
Leaf Movements
(LEINWEBER)

Temperature	Length of Period
°C	hr
15	28.3 ± 0.4
20	28.0 ± 0.4
25	28.0 ± 1.0

Table 4. Lizards (*Lacerta sicula*),
Running Activity
(HOFFMANN 1957)

Temperature	Length of Period
°C	hr
16	25.20
25	24.34
35	24.19

Of course, *the circadian oscillations can be used for time measurement only because of their surprisingly low dependence on temperature.* The synchronization of the oscillations with environmental cycles (as will be described) prevent even the transient errors already mentioned.

In lower organisms the degree to which the periods depend on the temperature may be influenced by the type of nutrition. Experiments in *Euglena gracilis* (rhythm in mobility in continuous darkness; BRINKMANN) revealed that in autotrophic cultures the frequency is independent of different values of constant temperature in the range of 15°C to 35°C. In mixotrophic cultures the frequency decreases with increasing values of constant temperature.

> To avoid misunderstandings, it should be remembered that the *amplitude* of a particular diurnal process may depend greatly on temperature. It is even possible that the processes controlled by temperature are limited by a minimum temperature. They may not function, or intermittently fail, at temperatures that still permit the clock to continue uninfluenced. In such a case the process that failed may become functional again without phase shift when the temperatures are restored to normal (see BÜNNING 1959).

Temperature coefficients less than 1. It is difficult to understand the surprising independence of temperature that usually occurs after

Table 5. Oedogonium, Sporulation
(Bühnemann)

Temperature	Length of Period
°C	hr
27.5	25
17.5	20

adapting the organisms for a few days to a particular temperature. There are several observations indicating an alternation of different processes with different temperature coefficients. At least they have been interpreted this way. One such observation concerns temperature coefficients less than 1—for example, the sporulation rhythm of the alga *Oedogonium* (Table 5), where a temperature coefficient (Q_{10}) of 0.8 was found. Also, the bioluminescence and cell division of *Gonyaulax* yielded values lower than 1.0 (0.85 to 0.9; Table 2).

Compensation by opposing processes? The change in period after lowering the temperature seems to indicate the importance of opposing processes with different temperature coefficients. Lowering the temperature can first cause a long delay as already mentioned, but after a few days it is followed by an acceleration, until finally the normal period reoccurs as a result of "adaptation." Thus, the independence of temperature is achieved. The diurnal leaf movements of beans may illustrate this. Plants usually showing a period of about 28 hours were transferred to a temperature of 15°C. The first period after the transfer was extended to 33.4 hours, but the consecutive periods were shortened to 26.2, 24.6, and 24.1 hours (Leinweber, see also Fig. 48). Larger temperature differences will intensify this effect. Similar relations were found with *Kalanchoe* flowers (Fig. 49). The influence of extremely low temperatures applied for a short time may produce such an instability. This rule holds, at least in many instances, for animals, too.

Observations like these may lead to the assumption that low temperature starts a compensatory process, becoming fully effective

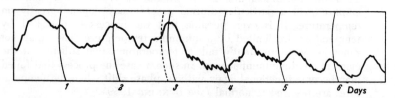

Fig. 48. *Phaseolus coccineus*, leaf movements in LL. Dotted line indicates change of temperature from 21°C to 10°C. The first period at 10° is longer, the following ones are shorter than at 21°. After Bünning u. Tazawa

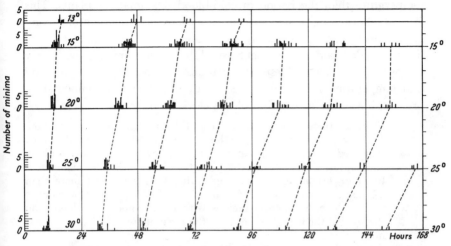

Fig. 49. *Kalanchoe blossfeldiana*. Time of minima of flower opening in DD at 15°, 20°, 25°, and 30°C. Ordinate: number of minima occurring at the indicated times. Abscissa: hours after the last 6-hour light period, preceding LL. Dotted lines connect median values. After OLTMANNS

after a certain time and finally even overshooting the mark; but some observations speak against it. We shall discuss this interpretation later. Such an exact compensation might be considered unnecessary for warm-blooded animals, yet they are also independent of temperature to a great extent. RAWSON carried out experiments with mice (*Peromyscus*) and hamsters. The body temperature was lowered for several hours to about +14°C; yet he found only a slight change of period length resulting in Q_{10} values between 1.05 and 1.3. And chilling bats down to about +11° yielded Q_{10} values of less than 1.1 throughout the temperature range used. Such a small influence of temperature was found also by MENAKER in the bat *Myotis lucifugus* for temperatures as low as 3° to 10°. FOLK obtained the same results by chilling ground squirrels to +5°. He recorded several diurnal functions (rectal temperature, heart frequency, oxygen uptake, and running activity). Warm-blooded animals also apparently show an errorless continuation of the clock, even during hibernation, when temperature is lowered. They tend to awake at a time of day characteristic of their normal activity phase (FOLK). Furthermore, isolated intestinal segments of hamsters are independent of temperature in the diurnal fluctuations of their contractions (BÜNNING 1958b). The circadian processes in isolated adrenal glands also show this temperature independence (ANDREWS and FOLK).

The period in warm-blooded animals is no more dependent on

temperature than the period in cold-blooded animals or plants. This is quite surprising since special provisions for temperature compensation by antagonistic processes appear to be unnecessary in warm-blooded animals (except for hibernating animals with low temperatures). Thus, these observations support the interpretation that temperature independence is not due to a special compensating process, but to the mechanism of the clock itself.

> There are reasons, however, to accept this conclusion with reservation. A study of the circadian rhythms of a number of tropical plants, examined by recording leaf movements, revealed a rather striking temperature dependence, even when transient cycles were eliminated. In *Phaseolus mungo*, for example, the periods reached more than 32 hours at 17°C (BÜNNING et al. 1966; MAYER). Temperature independence in tropical regions with little temperature fluctuation throughout the year does not have the great adaptive value that it has in other regions.

b. Temperature effects during different parts of the cycle

Normal physiological range. Here the discussion will be confined to temperatures within the "normal" range, that is within limits that permit oscillations of approximately normal periods. This is the above-mentioned temperature range, within which the length of period is largely independent of temperature. These temperatures are between +5° and +30°C for most objects.

Responses of the different phases. The relative independence of temperature of the period should not lead to the conclusion that the endogenous rhythm is insensitive to temperature. There must be some sensitivity since a diurnal temperature periodicity can entrain the rhythm to periods of exactly 24 hours in many plants and animals.

Testing the whole cycle by exposing only short sections of it to temperature changes revealed phases that are relatively insensitive to temperature and other phases that reacted quite strongly by lengthening or shortening the following periods. The leaf movements of *Phaseolus coccineus* demonstrate this quite well (Fig. 50). The plants were kept in LL at a constant temperature of 20°C. Increasing the temperature for 4 hours to 28° during or shortly after the lower position of the leaves considerably delayed the following lowest leaf position. The cycle was extended, but less than 4 hours later the system reacts exactly opposite, and the following lowest leaf position occurs several hours earlier. The cycle was shortened. The phase during which an increase of temperature causes the strongest advance occurs only 5

Temperature Effects

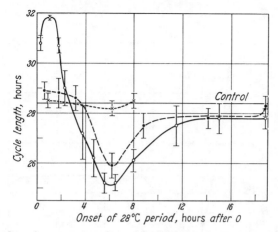

Fig. 50. *Phaseolus coccineus*, diurnal leaf movements in LL. Change of cycle length if temperature is raised from 20°C to 28°C for 4 hours. Abscissa: time of temperature change to 28°. Abscissa 0 refers to the maximum night position of the leaves. Control: cycle length in constant temperature of 20°C. o————o the directly influenced cycle. ●– – –● the following cycle. x- - - -x the third cycle. After MOSER

hours after the phase in which an increase in temperature results in maximum delay.

The curve depicting this kind of information is now generally called a *phase-response curve*. It indicates "how the amount and the sign of a phase shift, induced by a single . . . stimulus, depends on the phase in which the stimulus is applied" (*ASCHOFF *et al.*).

An increase in temperature affects not only the cycle immediately following the stimulus but also the cycle thereafter, as illustrated in Fig. 50. In this example, the final new phase position of the rhythm is reached only in the third cycle (calling the directly affected cycle the first cycle). This phenomenon of so-called *transients* is also known as a light effect. It was first described by PITTENDRIGH and BRUCE. Transients occurred in the aforementioned experiments only if an increase in temperature caused an acceleration in the running of the clock. They did not occur after a delay of the clock. Considering also the effects on the following cycles, which are influenced by the transients, a total phase shift results, as summarized in Fig. 51. This shift is the response to a single temporary increase of temperature.

Increasing the temperature not only temporarily, but also changing it from a constant temperature of 20°C to a constant temperature of 28° also shifts the phases in a manner that might be predicted on the basis of the results mentioned above (Fig. 52). Transients can be

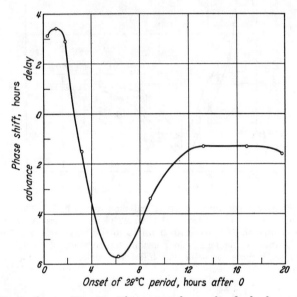

Fig. 51. Similar to Fig. 50. The curve shows the final phase shift if the temperature is raised from 20° to 28°C for 4 hours. After MOSER

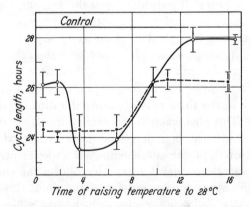

Fig. 52. Similar to Fig. 50, but the temperature remains 28° throughout the experiment, after being raised from 20° to 28°C. Abscissa: time of raising temperature to 28°C. o———o length of the cycle during which the higher temperature started. •– – –• length of the following cycle. After MOSER

distinguished here as well. The results showing specific responses of different phases make it easy to understand why transient cycles must occur after a transfer from one constant temperature to another, although the normal period appears again after a few cycles. The shortening may amount to more than 4 hours in the example of Fig. 50. Calculating these data results in Q_{10} values of 1.2 or more. This was the basis of earlier experiments, in which transients were not considered, it being assumed that the length of period is dependent on temperature to this extent (see Chapter 5(a)). Whether a change from one constant temperature to another constant temperature affects the periods with Q_{10} values somewhat larger or smaller than 1, after the transients are over, i.e., after the third or the fourth period, depends upon the quantitative proportions of the phases sensitive to temperature reacting antagonistically. The reaction of the phases to extreme low temperatures will be discussed later.

c. Setting the clock by temperature cycles

Effective temperature cycles. Through these studies of the effect of temperature, we understand that the clock can be set by temperature cycles. However, temperature cycles, repeated or single, are usually much less effective in setting the clock than LD cycles.

The leaf movements of *Phaseolus* can be entrained to a certain extent by diurnal changes between 25° and 30° (STERN u. BÜNNING). This means entraining the free-running periods of about 24 hours to exactly 24 hours. A change of 15° and 25° was tested successfully on the sporulation rhythm of the alga *Oedogonium* (BÜHNEMANN), but a temperature difference of 2.5° is equally effective. According to OLTMANNS, the diurnal petal movements of *Kalanchoe* are regulated by temperature fluctuations as small as 1°. ROBERTS was easily able to synchronize the activity rhythm of the cockroach *Leucophaea maderae* by temperature cycles with a maximum of 27° and a minimum of 22°. In lizards, temperature differences less than 1°C are sufficient for the entrainment (HOFFMANN 1969a, b). But in two species of field mice temperature cycles of 15:26°C did not cause a clear entrainment (HOFFMANN 1968). The activity of the snail *Agrolimax reticulatus* follows the diurnal change in temperature, and differences as small as 0.1°C are said to be effective (DAINTON).

In all cases checked carefully, the phase angle difference between the temperature cycle and the physiological cycle depends on the amplitude of the temperature cycle (POHL; HOFFMANN 1969a, b).

In general, temperature cycles with lower amplitudes seem to be more efficient in plants and in poikilothermic animals than in homeothermic animals (HOFFMANN 1969b). It is remarkable that there is actually an entraining influence in homeotherms. From this we must conclude that temperature influences are not always directly on the clock, but can be mediated by peripheral senses (ENRIGHT).

Type of control. According to all data available on plants and animals, cycles of high and low temperature are always regulated so that the phase of low temperature coincides with the physiological state usually reached during the night. This has been described for the diurnal leaf movements of beans (STERN u. BÜNNING), for the sporulation rhythm of *Oedogonium* (BÜHNEMANN), for the running activity of rats (CALHOUN), and for different insect species (BENTLEY et al.). For further examples (lizards, *Drosophila*, cockroaches, *Euglena*) see BRUCE.

Limits of entrainment. Temperature cycles, just as LD cycles, should not deviate too much from the 24-hour periodicity or the physiological rhythm will not follow the temperature rhythm. An 8:8 hour cycle of high and low temperature often has no synchronizing ("entraining") effect on plants and animals.

Temperature limits. The oscillator functions only within a certain range of temperature. This range depends on the extreme conditions to which the species is adapted. The clock of *Gonyaulax*, an alga of warmwater, fails if the temperature drops to 11.5°C. The alga still lives and shows luminescence at this temperature, but the diurnal fluctuations are no longer noticeable (HASTINGS and SWEENEY). The rhythms of algae from cooler regions continue at much lower temperatures (see Chapter 5(d)). Several species of vascular plants from the tropics may fail to oscillate at 12°C or less, whereas diurnal leaf movements in arctic plants can be observed at much lower temperatures (MAYER).

d. Influence of low temperature

Stopping the clock. The phases are shifted considerably if the organisms are cooled down to between 0° and +5°C. In some cases even 5° to 10° is sufficient. This has been described for plants and for animals, for the time sense of bees (KALMUS 1934; RENNER, Fig. 53), for the pigment migrations in the fiddler crab *Uca* (BROWN and WEBB; STEPHENS), for the movements of bean leaves (BÜNNING u. TAZAWA), and for the growth rhythms in the *Avena* coleoptile (BALL and DYKE). Temperatures of 4° to 5°C are sufficient to stop the clock in the spider *Arctosa perita* as judged by the time sense in orientation (PAPI et al.).

Temperature Effects

This effect is not connected with the misleading possibility that only the process controlled by the clock is influenced but not the clock itself. The induced phase shift is persistent after transferring the organisms from the low temperatures back to normal ones. Often the phase shift is so intense that the clock is delayed for the same length of time as the period of extreme cooling. We shall see, however, that the phase shift can even be longer than the cooling period.

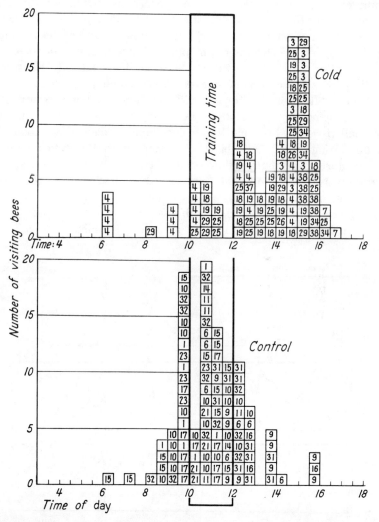

Fig. 53. Time sense in bees. For detailed explanation, see Fig. 1. Above: the bees were kept 5¾ hours at 4–5° after the last training period. Below: controls without cold treatment. After RENNER

Different temperatures are necessary for this kind of interference, depending upon the range of temperature to which the organism is adapted. Beans, for example, being adapted to relatively high temperatures, are retarded by little less than 10°C. The same goes for *Periplaneta americana*, a cockroach from warmer regions. Another cockroach, *Leucophaea maderae*, responds similarly, according to experiments by ROBERTS. In organisms that are usually also active at very low temperatures, this interference occurs only close to freezing point or below. Examples are the sporulation rhythm of the alga *Oedogonium* (RUDDAT) and the orientation rhythm of the pond skater *Velia currens*. When cooled to $+1°$ or $+2°$C this pond skater becomes motionless, yet the clock is not stopped (EMEIS). The pond skater is an animal preferring temperatures between 4° and 16°C (RENSING).

Responses of the different phases. More detailed experiments show that the oscillator is not "frozen solid" by these low temperatures, i.e., the clock is not simply arrested. The effect of low temperature limited to small parts of the cycle varies considerably depending upon the phase during which the object was exposed to the extremely low temperature (STEPHENS; BÜNNING u. TAZAWA; BÜNNING 1958a). In certain phases it has no or only little retarding effect, i.e., the following maximum of the tested physiological process occurs at the normal time. During other phases the rhythm is delayed by low temperature. This can be demonstrated in plants as well as in animals (Figs. 54 to

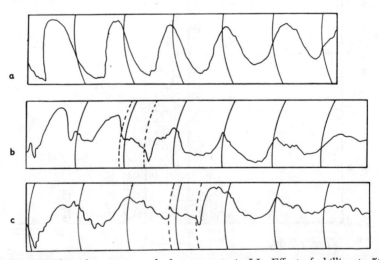

Fig. 54. *Phaseolus coccineus*, leaf movements in LL. Effect of chilling to 5°C during different parts of the cycle. Chilling period between broken lines. Guide lines 24 hours apart. Little delay in (b), strong delay in (c), control example without chilling in (a). After BÜNNING u. TAZAWA

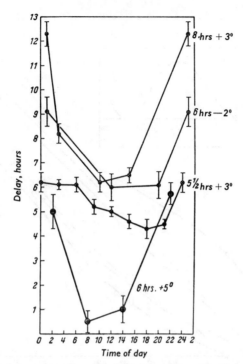

Fig. 55. *Periplaneta americana*. Rhythm of activity in LL (dim light). Abscissa: time of beginning low temperature treatment. Temperature and duration of low temperature are indicated. Vertical lines: standard deviation of mean values. Ordinate: delay of the activity peak during the ensuing days. After BÜNNING 1959

58, for additional experiments see WAGNER). Yet the delay does not necessarily last as long as the period of chilling. The oscillator is not always fixed in the phase in which it was caught by the low temperature. For example, cockroaches (*Periplaneta americana*) were cooled down just before reaching their maximum running activity. After returning them to normal temperature it takes more time to reach the maximum than would have been the case without the cold treatment. Often the delay is longer. Experiments by WILKINS on the rhythmic CO_2 output of *Bryophyllum* also show a similar response to chilling during different phases.

This kind of experiment strongly supports the hypothesis of oscillations including phases of relaxation. Cooling during the phases of tension lowers the oscillator to a level of less tension. After reestablishing normal temperatures, a phase shift is observed, not merely as long as the cooling period, but longer. This additional time is necessary to return the oscillator to the condition prior to the cold treatment. If the cooling period coincides with the relaxing phases, the situation is reversed,

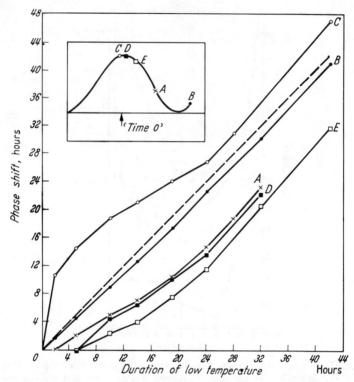

Fig. 56. *Phaseolus coccineus.* Diurnal leaf movements in LL and at 20°C. Influence of a single 5°C period. Abscissa: duration of exposure to this low temperature. Ordinate: phase shift due to this chilling. Beginning of low temperature treatment at different phases of the physiological rhythm, as indicated by A, B, C, D, E. Partial figure (top) shows positions of these phases. Time 0 corresponds to maximum physiological night position of the leaves. A is 8 hours after this zero value, B 16 hours after zero, C ½ hour before zero, D ½ hour after zero, E 3 hours after zero. Broken line shows phase shifts that should occur in case the low temperature should just stop the clock. After BÜNNING u. WAGNER (see WAGNER)

since relaxation can proceed quite easily, not requiring any additional energy.

As might be expected, after a very long cooling period, the oscillator apparently approaches more or less closely a certain final value, a "resting position" (Fig. 57). Raising the temperature again has a phase-setting effect. This can be deduced directly from curves recording leaf movements (Fig. 58). A corresponding response was found in cockroaches. BÜNNING (1958a) observed this in *Periplaneta americana* and ROBERTS found it in *Leucophaea maderae* for cold periods up to 57 hours. The crab *Carcinus maenas* shows the same effect: chilling to 4° for more than 6 hours resulted in a rephasing, with new peaks de-

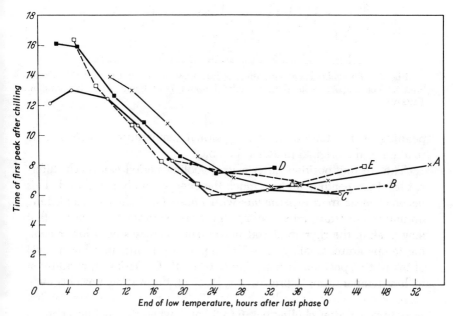

Fig. 57. Same experiments as in Fig. 56, but ordinate represents the time (hours) required from the end of chilling to the occurrence of the next phase 0 of the physiological cycle (for phase 0 compare top figure in Fig. 56). Moreover, the abscissa represents the end of the chilling periods. After BÜNNING u. WAGNER (see WAGNER)

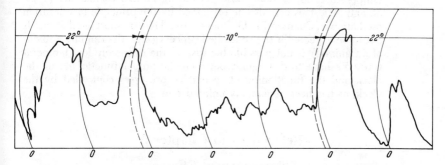

Fig. 58. Diurnal leaf movements of *Phaseolus coccineus*. Different temperatures as indicated in the figure; LL. Guide lines 24 hours apart. Note irregular cycles during 10° period and resetting the phase due to reestablishing normal temperature. After BÜNNING u. TAZAWA

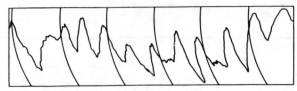

Fig. 59. *Phaseolus coccineus*, diurnal leaf movements in LL at 10°C. Guide lines 24 hours apart. Note short periods of about 7–14 hours. After BÜNNING u. TAZAWA

pending on the time of return to normal temperature; there was no relation to the original phases (NAYLOR).

The "resting position" after a long cooling period can reach different levels, depending on how much the energy supply is reduced. This becomes clear from experiments with *Phaseolus*. When reestablishing normal temperature after cooling, the time required to reach the first new peak in the rhythm of leaf movements increases with lower cooling temperature. Cooling to $+2°C$ apparently results in a lower level of the resting position than cooling to 6° or 10°C. Darkness, in addition to low temperature, depresses the level of the resting position even more. For example, according to Fig. 57, the time required for the first peak to occur after chilling was 6 to 8 hours when the cooling (here in light) was to $+5°C$. Additional experiments (BÜNNING 1963) proved it to be only 4 hours when the cooling (in light) was to 12°; however, it was nearly 12 hours when the cooling (in darkness) was to 6°.

> Sometimes short oscillations with a periods of approximately 7 to 14 hours can be observed in a transitional temperature range, i.e., at temperatures lower than those permitting normal oscillations (Figs. 58 and 59). This can also be explained by the principle of relaxation oscillations (Chapter 7). Of course, this kind of short oscillations is not to be expected if the controlled process (e.g., the running activity of an animal) is not possible because of the extremely low temperature. We have to distinguish between the proper functioning of the clock and the functioning of secondary processes controlled by the clock, as has been repeatedly pointed out.

References to Chapter 5

(*Temperature Effects*)

a. Reviews

°ASCHOFF, J., K. KLOTTER, and R. WEVER. 1965. Circadian vocabulary. In *Circadian Clocks*. J. ASCHOFF, ed. Amsterdam: North Holland Publishing Company. Pp. X–XVIII.

*SWEENEY, B., and J. W. HASTINGS. 1960. Effects of temperature upon diurnal rhythms. Cold Spring Harbor Symp. quant. Biol. 25:87–104.
*WILKINS, M. B. 1965. The influence of temperature and temperature changes on biological clocks. In *Circadian Clocks*. J. ASCHOFF, ed. Amsterdam: North Holland Publishing Company. Pp. 146–163.

b. Other references

ANDREWS, R. V., and G. E. FOLK. 1964. *Comp. Bioch. Physiol.* 11:393–409.
BALL, N. G., and I. J. DYKE. 1964. *J. exp. Bot.* 5:421–433.–
BENTLEY, E. W., D. L. DARIN, and D. W. EWER. 1942. *J. exp. Biol.* 18:182–195.–
BRINKMANN, K. 1966. *Planta (Berl.)* 70:344–389.–
BROWN, F. A., and H. M. WEBB. 1948. *Physiol. Zool.* 21:371–381.–
BRUCE, V. G. 1960. *Cold Spring Harbor Symp. quant. Biol.* 25:29–48.–
HASTINGS, J. W., and B. M. SWEENEY. 1957. *Proc. Nat. Acad. Sci. USA* 42:676–682.–
BÜHNEMANN, F. 1955. *Z. Naturforsch.* 10b:305–310.–
BÜNNING, E. 1931. *Jahrb. wiss. Bot.* 75:439–480;–
———. 1935. *Ber. dtsch. Bot. Ges.* 53:594–023;–
———. 1958a. *Biol. Zentralbl.* 77:141–152;–
———. 1958b. *Naturwiss.* 45:68;–
———. 1959. *Z. Naturforsch.* 14b:1–4;–
———. 1963. *Z. Bot.* 51:174–178.–
BÜNNING, E., G. HAILER, u. W. MAYER. 1966. *Ber. dtsch. Bot. Ges.* 79:7–14.–
BÜNNING, E., u. M. TAZAWA. 1957. *Planta (Berl.)* 50:107–121.
CALHOUN, J. B. 1945. *Ecology* 26:250–273.
DAINTON, B. H. 1954. *J. exp. Biol.* 31:165–187.
EMEIS, D. 1959. *Z. Tierpsychol.* 16:129–154.–
ENRIGHT, J. T. 1966. *Comp. Bioch. Physiol.* 18:463–475.
FOLK, G. E. 1957. *Amer. Naturalist* 91:153–166.
HASTINGS, J. W., and B. M. SWEENEY. 1957. *Proc. Nat. Acad. Sci. USA* 43:804–811.
———. 1958. *Biol. Bull.* 115:440–458.–
HOFFMANN, K. 1957. *Naturwiss.* 44:358;–
———. 1963. *Z. Naturforsch.* 18b:154–157;–
———. 1968. *Verhandl. Deutsch. Zool. Ges.* 265–274, Leipzig, Akad. Verl. Ges;–
———. 1969a. *Z. vergl. Physiol.* 62:93–110;–
———. 1969b. *Oecologia* 3:184–206.
KALMUS, H. 1934. *Z. vergl. Physiol.* 30:405–419;–
———. 1935. *Biol. Gen.* 11:93–114.
LEINWEBER, F. -J. 1956. *Z. Bot.* 44:337–364.
MAYER, W. 1966. *Planta (Berl.)* 70:237–256.–
MENAKER, M. 1961. *J. cell. comp. Physiol.* 57:81–86.–
MOSER, I. 1962. *Planta (Berl.)* 58:199–219.

NAYLOR, E. 1963. *J. exp. Biol.* **40**:669–679.
OLTMANNS, O. 1960. *Planta (Berl.)* **54**:233–264.
PAPI, F., L. SERRETTI, u. S. PARRINI. 1957. *Z. vergl. Physiol.* **39**:531–561.–
PITTENDRIGH, C. S. 1954. *Proc. Nat. Acad. Sci. USA* **40**:1018–1029.–
PITTENDRIGH, C. S., and V. G. BRUCE. 1959. In *Photoperiodism and Related Phenomena in Plants and Animals.* R. B. WITHROW, ed. Pp. 475–505. Washington: Am. Ass. Adv. Sci.
POHL, H. 1968. *Z. vergl. Physiol.* **58**:364–380.
RAWSON, K. S. 1956. Ph.D. thesis. Harvard University.
——. 1960. *Cold Spring Harbor Symp. quant. Biol.* **25**:105–113.–
RENNER, M. 1957. *Z. vergl. Physiol.* **40**:85–118.–
RENSING, L. 1961. *Z. vergl. Physiol.* **44**:292–322.–
ROBERTS, S. K. 1962. *J. cell. comp. Physiol.* **59**:175–186.–
RUDDAT, M. 1961. *Z. Bot.* **49**:23–46.
STEPHENS, G. C. 1957. *Physiol. Zool.* **30**:55–69.–
STERN, K., u. E. BÜNNING. 1929. *Ber. dtsch. Bot. Ges.* **47**:565–584.–
SWEENEY, B. M., and J. W. HASTINGS. 1960. *Cold Spring Harbor Symp. quant. Biol.* **25**:87–104.
WAGNER, R. 1963. *Z. Bot.* **51**:179–204.–
WAHL, O. 1932. *Z. vergl. Physiol.* **16**:529–589.–
WILKINS, M. B. 1962. *Proc. roy. Soc. B.* **156**:220–241.

6. Light Effects

Die periodische Bewegung an sich ist unabhängig von dem Wechsel der Beleuchtung, aber die periodische Bewegung in dem Zeitmaß, wie sie unter gewöhnlichen Verhältnissen auftritt, wird durch den Lichtreiz bestimmt. . . .

J. SACHS, Flora Nr. 30, Regensburg 1863, p. 469.

(Periodic leaf movements can continue without LD cycles, but light controls the phases and periods.)

a. Effects of continuous light

Length of period in LL and DD. Normal LD cycles usually entrain (synchronize) the circadian periods to periods of exactly 24 hours. LL modifies the length of period. This effect has been repeatedly described since 1939 (JOHNSON).

Bullfinches kept in DD show a period of 24 hours; in LL this period changes to 22 hours (ASCHOFF 1953). In mice, however, the period amounts to 25 or 26 hours in LL, but is reduced to 23 or 23.5 hours in DD (ASCHOFF 1955b; MEYER-LOHMANN). According to HEMMINGSEN and KRARUP (1937) and to BROWN et al. (1956), the period in rats is 26 hours in LL and 24 hours in DD. PITTENDRIGH and BRUCE (1957) reported a lengthening of the periods by LL in a mouse species (*Peromyscus maniculatus rufinus*), but the emergence rhythm of *Drosophila* shows shorter periods in light (BRUCE and PITTENDRIGH). According to *ASCHOFF (1960), the frequency changes linearly with the logarithm of the light intensity (Fig. 60). ASCHOFF expressed a general rule ("circadian rule"): the length of period of animals active in light decreases with increasing light intensity; and in animals active during darkness it increases with increasing light intensity (Fig. 60). There are, however, exceptions to this rule (see *HOFFMANN; RENSING u. BRUNKEN; POHL 1972; CAIN and WILSON; MARTINEZ).

These statements refer to experiments with white light. It is conceivable that the contrasting responses of the different animal species are to a certain degree due to a different mode of action of certain colors. In plants, certainly, an antagonistic effect of red and far-red

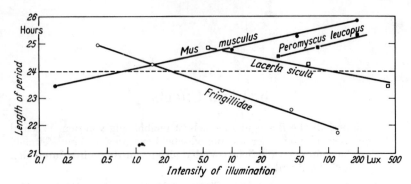

Fig. 60. Spontaneous period of animals in constant environment, depending on intensity of illumination (LL). *Peromyscus leucopus* according to JOHNSON, *Lacerta sicula* according to HOFFMANN. After ASCHOFF 1959

light was found. The leaf movements of beans (*Phaseolus coccineus*) show different periods, depending on the quality of light (see Table 6).

A dependence on the quality of light was also found in circadian leaf movements of *Coleus* (HALABAN).

It is remarkable that not only are the periods in far-red light shortened, but also the rhythm fades out much quicker than in DD or in red light (Fig. 61). Far-red light results in irregular short cycles. We have already seen a similar phenomenon when studying the influence of low temperatures. When a factor produces short periods and also causes the rhythm to fade out quickly, it suppresses one part of the oscillation before the normal final value is reached. The opposite part of the oscillation is thereby also shortened, thereby shortening the oscillation as a whole.

In general, the differences between length of period in DD and LL amount to only a few hours or less. This finding parallels the observation of approximately equal periods at different constant temperatures.

Table 6. *Phaseolus coccineus*, Leaf Movements.
Quality of Light and the Length of Periods (LÖRCHER)

Light Condition	Length of Period
LL, nm	hr
DD	26.3 ± 0.2
340–450	26.6 ± 0.3
450–610	26.2 ± 0.4
610–690	28.1 ± 0.2
690–850	24.7 ± 0.3
Longer than 850	26.0 ± 0.4

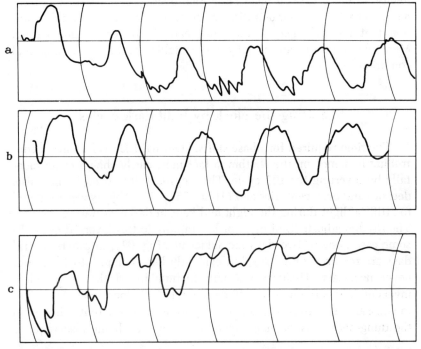

Fig. 61. *Phaseolus coccineus*, leaf movements. (a) DD, (b) LL, red (610–690 nm), note longer periods, (c) far-red (>700 nm), note depression of rhythm. Guide lines 24 hours apart. After LÖRCHER

Like temperature, however, light strongly influences the phase position of the rhythm. In discussing the influence of temperature, a solution was found to this seeming discrepancy. Also, with light there are phases reacting antagonistically, i.e., phases reacting to light by delaying the following part of the oscillation, and other phases reacting in an opposite way. Depending on the relative proportions of these two effects, LL may result in a lengthening or a shortening of the periods. For that reason we will have to return to the "circadian rule" after describing the response curves.

Fade-out in continuous light. After seeing that one light quality favors the continuation of the rhythm for many days while another one suppresses it, it is not surprising to find so many diverse statements about the advantages of DD or LL in maintaining the rhythm. Usually, DD (or LL of dim light) is a better condition for plants or animals than LL with high intensity, but the reverse can also be true.

Of course, demonstrating a fade-out of the overt rhythm in LL or DD does not necessarily indicate a fade-out of the underlying rhyth-

92 The Physiological Clock

micity. In many cases it means only that overt processes, such as locomotor activity, are so much directly disturbed by the abnormal condition that the influence of the clock can no longer be dominating. Metabolic rhythms are usually more stable under these abnormal conditions.

b. Setting the clock by light-dark cycles

The time required for phase shifts. Diurnal LD cycles have a controlling influence on the rhythm, a fact most easily shown experimentally by inverting the LD cycle. This approach was tried by botanists decades and even centuries ago (HILL 1757). The objects were exposed to artificial light during the night and kept in darkness during the day. The rhythm adjusts itself more or less rapidly to this inversed cycle. In some cases, one or two days are sufficient (Fig. 62); in others several days are required (Fig. 63), but evidently longer than 8 to 16 days is never necessary. HEMMINGSEN and KRARUP found that a complete inversion in rats requires 8 to 10 days, a relatively long time. According to GEISLER, 11 to 16 days are necessary to invert the activity rhythm in the dung-beetle (*Geotrupes silvaticus*). A human being also requires

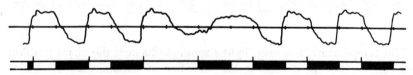

Fig. 62. *Chenopodium amaranticolor*, diurnal leaf movements. Inversion of the LD cycle during the third period. Guide marks 24 hours apart. Dark periods indicated by black lines. The physiological rhythms is completely rephased within two days. After KÖNTIZ (original)

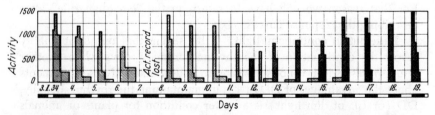

Fig. 63. Spontaneous muscular activity of a female albino rat before and after reversal of light and dark. Dark periods indicated by black lines. Reversal of the LD cycle on Jan. 9. Complete reversal of the physiological rhythm on Jan. 17. Ordinate: revolutions of activity wheel per hour. After HEMMINGSEN and KRARUP

Light Effects

that much time when flying east or west from one continent to another. For other examples and general discussion, see Aschoff u. Wever (1963).

In human beings, the course of the body temperature does not invert at once when the pattern of living has been reversed (e.g., due to night work). This rhythm takes more than one week for a complete reversal (see Aschoff 1955b). Shifting the LD cycle and the meal times by 12 hours results in a shift of the phase of the temperature curve by only 6 to 10 hours after one week. Usually the inversion is complete after 9 to 10 days. Experiments in the arctic (Sharp) and during flights (Burton; LaFontaine et al.) also showed that it takes several days for a complete phase shift in humans. This adjustment is faster during delaying changes (east-to-west flights) than it is during advancing changes (west-to-east flights; Klein et al).

Basically, the same new setting of the phases occurs if, instead of an inverse LD cycle, a slightly shifted one is given—for example, a cycle shifted by 6 hours in relation to the original. In some cases, the complete inversion—i.e., a phase shift of 180°—has an immediate effect, whereas a phase shift of a few hours may synchronize the physiological rhythm only after several days (see also Pittendrigh 1960). One likelihood is that an inversion of the rhythm (180° phase shift) may be of the nature of a "total reset," whereas a change of a few hours may involve a genuine phase shift with the attendant transients.

As might be expected, the time required for synchronization is shorter with higher light intensities than with lower intensities (Wobus).

LD cycles are the most effective synchronizers for circadian rhythms. Only rarely are LD cycles less effective than temperature cycles and rarely does light have no influence at all (Tweedy and Stephen, the discussion on circatidal rhythms, Chapter 11).

Phase angle differences. The peaks and troughs of the several physiological rhythms show quite different intervals of time from the peaks and troughs (or other phases) of the synchronizing LD cycles. Those phase angle differences also depend on the difference between the period of the physiological circadian rhythm and the entraining period (Aschoff u. Wever 1962; see also the remarks in Chapter 14).

The length of the light period also exerts some influence on the phase angle difference (Fig. 64). The same dependence becomes evident by observing the behavior of an organism over the period of a year. Under the influence of changing day lengths, the physiological phase drifts in a pattern which does not show any precisely fixed time relation to either dawn or dusk.

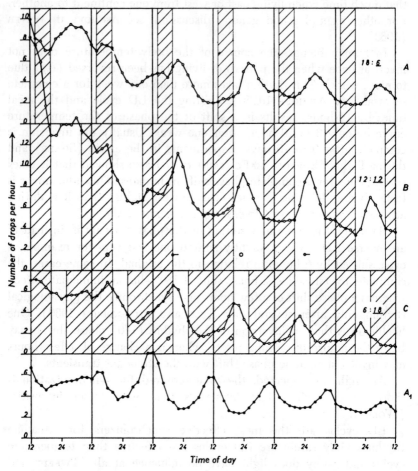

Fig. 64. *Amaranthus caudatus*. Periodicity of bleeding during 24-hour LD cycles. Length of light period: 18 hours (A), 12 hours (B), 6 hours (C). Beginning of the light period always at 20.00 hours (P.M.). (A_1): dark control. After HASSBARGEN

The annual changes in the duration of twilight have an influence on these changes of the phase angle differences too (WEVER; *ASCHOFF 1969).

Intensities required for entrainment and phase shifts. Since circadian rhythms are used for time measurement, a high accuracy in the running of these clocks is necessary: entrainment to exactly 24-hour cycles and a phase angle difference between the circadian rhythm and the normal LD cycle that does not vary from day to day. This cannot be reached by taking the light intensities during sunrise or sunset as

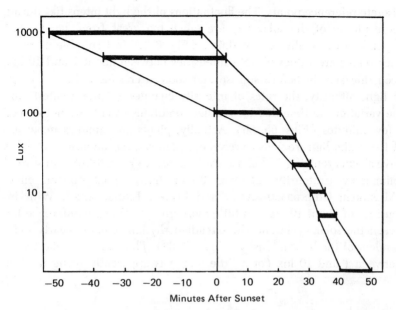

Fig. 65. Intensities of zenith-light, reaching a horizontal plane. Measured in Tübingen, March 1–12. Note that the rate of change of light intensities is greatest when the intensity has reached about 10 lux after sunset. Arbitrary variations due to cloudiness have their minimum in these regions of twilight. From BÜNNING 1969

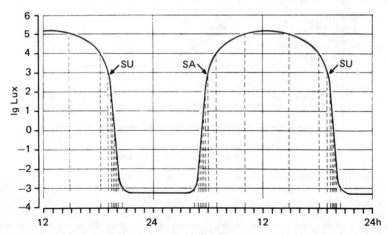

Fig. 66. Diurnal changes of light intensity at Tübingen on April 2, 1965. Abscissa: time of day. Ordinate: light intensity, lg lux. SU: sunset, SA: sunrise. The curve shows (just as Fig. 65 does) that intensities between about 1 and 10 lux, reached about half an hour after sunset and about half an hour before sunrise, are the best reference values for a physiological process. From ERKERT

discrete reference points. The fluctuations of the light intensities during these phases of the solar day are much too great from day to day, depending especially on cloudiness (Fig. 65). Most suitable as reference points are values of light intensities between about 1 and 10 lux, occurring shortly before sunrise and soon after sunset. In this range of light intensity, the rates of intensity changes are the greatest, and the variations in time due to weather conditions usually do not exceed a few minutes (Figs. 65, 66). Actually, plants and animals make use of these dim light regions as reference points. Even intensities of 1 lux have a synchronizing and phase-shifting effect (Figs. 67, 68); in several animals even intensities of about 0.1 lux have a phase-shifting effect (MENAKER; CHANDRASHEKARAN and LOHER; ENGELMANN). With intensities of about 10 lux or a little more, the phase angle difference between the solar cycle and the circadian rhythm remains nearly independent of the light intensity (Figs. 67, 68). Thus, the very short span between 1 and 10 lux (or a little more) is apparently being used as

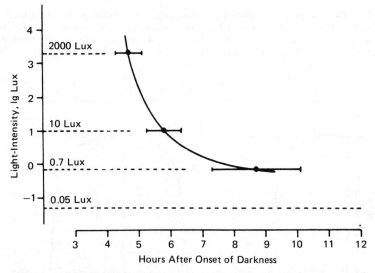

Fig. 67. *Glycine max* (soybean). Circadian leaf movements in 12:12 hours LD cycles. Abscissa: appearance of night peaks, hours after the beginning of the daily light period. There was no synchronizing effect in case the light intensity was only 0.05 lux. With this extremely low intensity, the light-dark cycle acted like DD and resulted in a free-running period of more than 26 hours (this is not shown in the figure). Intensities of 0.7 lux or more synchronized the circadian cycle to 24-hour cycles. The figure shows that with intensities of 10 lux the phase angle difference between the LD cycle and the circadian rhythm is already nearly the same as for LD cycles with much higher light intensities. Data from BÜNNING 1969

reference point. In addition, an abrupt increase (or decrease) in intensity (Figs. 65, 66) is, of course, more effective in controlling a physiological process than a gradual one.

Under laboratory conditions, the susceptibility to resetting the phases by light proved to be dependent on the pretreatment. In *Drosophila*, after three days of DD, the sensitivity increased more than tenfold (WINFREE).

Organs perceiving the controlling light stimulus. In some cases, light directly affects the cells that exhibit periodicity. This phenomenon is obvious in unicellular organisms, but it also occurs in different tissues and organs of higher plants.

In mammals, periodicity is usually controlled via the eyes, so that the absence of eyes or covering the eyes results in the free-running of the rhythm with typical individual deviations from 24-hour periodicity. The control mediated by the eyes is exerted over the adrenal cycle, mentioned earlier, and all the functions connected with it (HALBERG et al.).

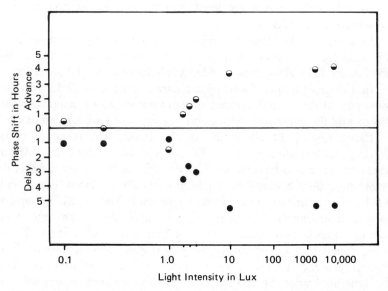

Fig. 68. The extent of advancing and delaying phase shifts effected in the circadian system of *Drosophila* by light pulses of 15-minute duration and varying intensities.

The circles denote the calculated median hours for the eclosion peaks of experimental populations on days 4 and 5 after exposure to light pulses. The line separating advances from delays indicates the position of the eclosion peak medians of the control populations which did not receive any light signal. From CHANDRASHEKARAN and LOHER

In experiments with cockroaches by ROBERTS, phase shifts of the locomotor rhythm by light were not possible when only the ocelli were exposed and the compound eyes were covered. According to this finding, the compound eyes, not the ocelli, are the perceptive organs for phase-shifting light stimuli. This is also true for the stridulatory rhythm of the cricket *Teleogryllus commodus* (LOHER). In other insects, ocelli as well as compound eyes are active in mediating phase-shifting light signals (NOWOSIELSKI and PATTON).

But apparently, a light reception effective in phase shifts and entrainment is also possible without the eyes. CLOUDSLEY-THOMPSON reported this for the cockroach *Periplaneta*, and PAGE and LARIMER reported it for the crayfish *Procambarus clarkii*.

For insects, as well as for vertebrates, direct light effects on cephalic regions not including the eyes are known. The effects include phase shifts and entrainment to the exact 24-hour rhythm. In *Drosophila melanogaster*, LD cycles are effective in entraining the eclosion rhythm in an eyeless mutant (ENGELMANN u. HONEGGER).

The photoreceptive pigment, however, is located in the brain even in nonmutant *Drosophila*. The pigment is not the same as the one involved in vision and is not located in the compound eyes or ocelli (ZIMMERMAN and IVES).

In species of the katydid *Ephippiger*, compound eyes and ocelli are the main light receptors. But entrainment and phase shifting is also possible by light absorption in other cephalic regions (DUMORTIER).

In the grasshopper *Chorthippus curtipennis*, even light reaching some part of the central nervous system through an extracephalic path can control the rhythm of oviposition (LOHER and CHANDRASHEKARAN).

Extraretinal light absorption in the brain with resultant phase shifting and entrainment is known to occur in amphibians (*ADLER; ADLER; TAYLOR and FERGUSON). It is remarkable that even in birds the extraretinal light absorption can fully entrain with intensities of only 0.1 lux (experiments with blinded sparrows, MENAKER). In several species of mammals, especially in young individuals, the light penetrating through the ossified skulls is sufficient for such effects (BRUNT et al.; ZWEIG et al.). Apparently, the pineal gland can mediate the light cues in certain cases. But there are species of birds and fishes in which the synchronization by LD cycles is possible without participation of the eyes and the pineal gland (ERIKSON). Pinealectomized white-throated sparrows (*Zonotrichia albicollis*) are synchronized by LD cycles. But their activity rhythm decays to arrhythmicity in LL (MCMILLAN).

Involved pigments. The absorbing pigments responsible for the phase shifts can vary considerably from case to case, according to available experimental results. In fungi only blue light and ultraviolet are effective, which hints that flavins or carotinoids are the decisive pigments (SARGENT and BRIGGS). In several higher plants, long-wave irradiation is much more effective than short-wave irradiation (see LÖRCHER for *Phaseolus,* WILKINS for *Bryophyllum,* and HOLDSWORTH for *Bauhinia*). This hints to a possible role of the phytochrome. But in the case of other plant species, light absorption in chlorophyll seems to be decisive (KARVE and JIGAJINNI).

In insects, only short wavelengths up to about 500 nm seem to be effective (BRUCE and MINIS). FRANK and ZIMMERMAN, working with *Drosophila,* found no difference between the action spectra for delay and advance. However, in plants the action spectra for advance and delay can differ. In *Coleus* (circadian leaf movements), red light was effective in advancing the phases, and blue light in delaying the phases (HALABAN).

These examples should suffice to show that action spectra of phase shifts by light do not permit any conclusions about the nature of the substances participating in the clock. Apparently, the clock may be coupled to quite different photoreceptors. There may even be no coupling to a photoreceptor. For instance, the emergence rhythm in the leafcutter bee, *Megachile rotundata,* can be synchronized by a temperature pulse but not by LD cycles (TWEEDY and STEPHEN). Tidal rhythms might be other examples of circadian rhythms without coupling to photoreceptors (Chapter 11).

The influence of ultraviolet, as found by EHRET in *Paramecium,* might be an effect on the clock itself. The same conclusion may be drawn from SWEENEY's experiments on phase shifts in *Gonyaulax* (rhythms of luminescence) by ultraviolet light. The UV effects differed in several respects from effects of visible light. These results lead the author to the conclusion that UV acts directly on proteins or nucleic acids which thus must form part of the clock machinery.

Rapidity of influence on different functions. It was mentioned earlier that not all periodic changes in higher organisms have to be subjected to a common control. The unequal effect of light on individual rhythms is an expression of this. A change in the controlling LD cycle may shift some of the diurnal functions quicker than others. This is also known for humans. Some functions adjust themselves 2 or 3 days after reversing the LD cycle, whereas other functions adjust themselves only after 8 or 10 days to the new LD cycle. For example, inverting the rhythm of water excretion may require this time (SHARP). Shifting the phases at unequal rates implies that the functions can be dissociated. Such dissociations of periods ("disphasia") have been described often. In starlings, a 6-hour phase shift of the LD cycle synchronizes the activity rhythm within 2 days, but the internal clock used for orientation is adjusted only after 4 days (more examples are given by Aschoff 1958; Tongiorgi; Schmidt-Koenig, see also Chapter 14).

Return to the old phase position. It is almost self-evident that the periodicity will continue with the shifted phase after the rhythm has been inverted and the plant or animal has been transferred to constant conditions (i.e., to LL or DD). Nevertheless, there are some exceptions. Brehm u. Hempel report that the activity rhythm of the Colorado potato beetle can be reversed within 3 days, but in DD it returns to the normal phase positions within 48 hours.

> Exceptions such as this (more are mentioned by Aschoff 1954) permit several conclusions. First of all, one should look for unnoticed synchronizers. But the following seems more important: some endodiurnal functions in a multicellular organism can be shifted more easily than others. Thus, these functions are temporarily out of synchronization in the normal physiological sense. Perhaps the rhythm of the running activity is shifted rather early, while the rhythm of the digestive tract remains unchanged until later (corresponding to observations with humans during plane trips). If the synchronizer for the running pattern is excluded at this point, the periodicity of the other functions may act as a synchronizer for the running activity, i.e., its periodicity is forced back into the old phase position. This would also explain the observation that in some cases the back shifting may proceed faster than the primary shift after the original LD cycle had been reestablished (see Schmidt-Koenig; Geisler). Similar alterations in body temperature cycles were observed in humans during flights; although 3–5 days were required for primary phase shifting, only 1 day was needed for reversal (Hauty and Adams).

These facts indicate the existence of several oscillators affecting each other. On this basis we may also be able to understand why phase shifts may proceed more slowly under a new LD regime when the organism is more developed. More oscillators may be linked in vertebrates than in lower animals or plants. Furthermore, these links may also explain why a phase shift in the LD cycle results in relatively small physiological phase shifts within the first days. The shifts are quicker later on, as was reported by Halberg et al. (1960) for the shifting of the liver-glycogen rhythm in mice.

Rules of control by light; phase-response curves. An altered LD cycle causes new phase positions, as can be studied by experiments in which the phase shift is accomplished by short light perturbations given at various phases within the cycle. The light perturbations do not have to be very long to result in a phase shift. It is sufficient to expose plants or animals each day at the same time to a light period of a few hours, often only of a few minutes or even less, thereby interrupting

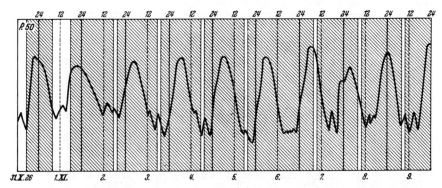

Fig. 69. *Canavalia ensiformis*, diurnal leaf movements. Dark periods shaded. Two-hour light periods offered from 17.30 to 19.30 hours induce delay (phase shift to the right). Beginning Nov. 7, light breaks were offered from 8–10 hours. These signals induce advance (phase shift to the left). After KLEINHOONTE

DD. KLEINHOONTE (1929) observed this in the leaf movements of *Canavalia* (Fig. 69, see also FLÜGEL) and PITTENDRIGH in the diurnal fluctuations of *Drosophila* emergence. The rhythm is shifted, then, in such a way that within a few days the phases normally associated with the day coincide with these light periods.

Investigations by BROWN et al. (1954) on the change in color of the fiddler crab *Uca* and also the experiments by HASTINGS and SWEENEY on the rhythmical luminescence of *Gonyaulax* are good examples. The light period does not have to last for several hours, but only for a few minutes. Often, fractions of a second of light of high intensity are sufficient to obtain synchronization by LD cycles. BRUCE et al. (1960) found a light flash of $1/2000$ second to be enough to trigger the discharge of sporangia in the fungus *Pilobolus sphaerosporus*. But even in these cases in which a single flash can completely reset the phases, the shifting does not take place immediately but only over several days, i.e., the so-called "transients" occur (PITTENDRIGH et al. 1958; Fig. 75).

Judging by evidence now available, light signals interrupting DD (or a temporary increase of light intensity during continuous dim light) have the same effect as an increase of temperature. This was investigated in some detail in the leaf movements of *Phaseolus coccineus*. The correspondence extends so far that phases responding to a light signal by maximum delay are identical with phases responding to an increase in temperature by maximum delay. And phases during which an increase in temperature causes maximum acceleration of the clock respond the same way to a light stimulus. This coincidence can be

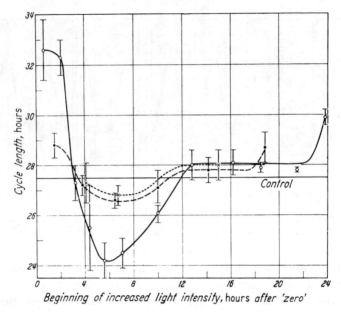

Fig. 70. *Phaseolus coccineus*, leaf movements in LL. Change of cycle length if light intensity is increased from 100 to 15,000 lux for 3 hours. Abscissa: time of intensity change to 15,000 lux. Abscissa 0 refers to the maximum night position of the leaves (cf. Fig. 56). Control: cycle length in LL, intensity 150 lux. o———o the directly influenced cycle, ●– – –● the following cycle, x- - - -x the third cycle. After MOSER

deduced by comparing Figs. 70 and 50. These figures also show the occurrence of transients in both cases. Another analogy may be pointed out. Instead of a temporary light signal, one can change from DD to LL, or from LL of dim light to LL with higher intensity. Phase shifts will then occur, which would be expected on the basis of the different sensitivity during individual phases, as was pointed out earlier. This corresponds to changing from one level of constant temperature to another level (see Figs. 71, 52). Lowering the light intensity results in the expected antagonistic responses.

In the object just described, the phase responding with maximum acceleration is separated by several hours from the phase responding in an opposite way. In some other objects the two phases follow each other very closely, practically without transition (Figs. 72, 73). The emergence rhythm of *Drosophila* also reacts in this latter way PITTENDRIGH 1960).

There are differences with respect to the transients. In the case of *Phaseolus* strong transients become evident only as shortened cycles,

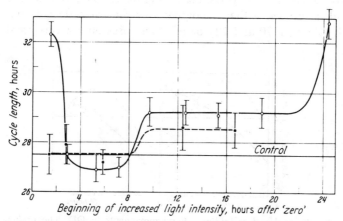

Fig. 71. *Phaseolus coccineus.* Similar to Fig. 70, but the light intensity remains at 15,000 lux throughout the experiment after being raised from 100 lux to 15,000 lux. Abscissa: time of raising light intensity. o——o length of the cycle during which the higher light intensity started ●---● length of the following cycle. After MOSER

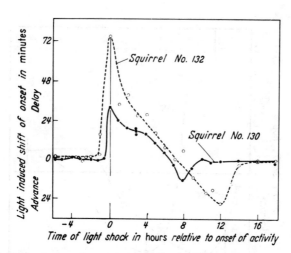

Fig. 72. Daily rhythm of resetting by light for 2 flying squirrels, Nos. 130 and 132. Phase shifts of the activity rhythms induced by 10-minute light shocks during DD are graphed, giving the number of hours before or after onset of activity when the light shock started. After DECOURSEY 1961

i.e., when the clock is advanced. This seems to be a rather general rule (*ASCHOFF 1965; Fig. 75). In other cases, transients may also be expressed by lengthened cycles, i.e. during delays. Different objects vary in another respect. Either major parts of the cycle do not show any sensitivity (Figs. 70 and 72) or in other cases only a very small part

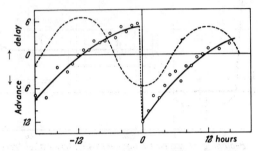

Fig. 73. *Kalanchoe blossfeldiana.* Petal movements in DD. The broken curve shows the movements of the controls without light breaks. Time 0 is by definition the time of complete closure of the flower. It is also the phase that separates delay responses from advance responses. This zero phase occurs in normal LD cycles during the dark period. Compare with Fig. 3. Circles indicate phase shifts after a single exposure to 2 hours of orange light. Abscissa: beginning of these light breaks. Ordinate: the phase shift in hours. After ZIMMER

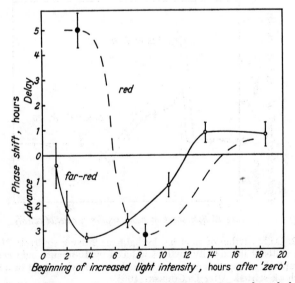

Fig. 74. *Phaseolus coccineus.* Similar to Fig. 70. LL, white light was not supplemented by a 3-hour period of higher intensity of white light, but by a 3.5 hour-period of red or far-red light. Compare with Fig. 70. Note the great difference in response curves. After BÜNNING u. MOSER

Light Effects

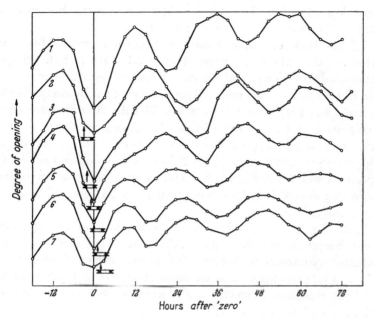

Fig. 75. *Kalanchoe blossfeldiana.* Petal movement in DD. For time 0 see Fig. 73. The curves show phase shifts after a single exposure to 2 hours of orange light. Time of these signals indicated by symbols in which the arrows mark the beginning. Curve 1: control without light break. Note shortened transients. After Zimmer

of the cycle is insensitive (Figs. 73, 74, 75). But these differences might be only quantitative differences. It is conceivable that the very same object would respond differently to different light intensities. The dark periods of a controlling LD cycle usually must last at least 3 to 8 hours. Shorter dark periods rarely suffice (Roberts 1959).

Variability of response curves. Sometimes rather far-reaching conclusions are drawn from response curves, especially with respect to the nature of the oscillator. These conclusions must be considered with reservation. The effect of light is, perhaps with the exception of ultraviolet, a very indirect effect that is mediated by different types of light receptors. This is true even for plants. In *Phaseolus* the response curves shown in Fig. 70 are predominantly due to the red component (about 650 nm) of the white light, and this light induces those phase shifts only when absorbed in the leaf joints themselves, i.e., in the tissues that enable the movements by virtue of their turgor changes. Far-red light (longer than 700 nm) is also active, and its effects lead to quite a different response curve (Fig. 74). Far-red light acts only when absorbed by the blade, not when absorbed in the leaf joint. In *Coleus*, a strong dependence on the quality of light must account for the difference in action spectra for advance and delay.

Relation between response curve and "circadian rule." In LL the period is either shorter or longer than in DD (see p. 89). The proportions between delays and advances found in response curves are a main factor for these differences. This becomes clear by comparing the red and far-red response curves of Fig. 74 with the respective effects of continuous red or far-red (Table 6). Far-red, causing a response curve with dominating advances (i.e., abbreviations of the cycle), induces rather short periods (24 to 25 hours), when offered as LL. The opposite is true for the red light, which causes response curves with dominating delays. It induces periods of about 28 hours when offered continuously.

Thus, the "circadian rule" and its "violations" do not necessarily reveal characteristics only of the oscillator itself, but also of the light-mediating physiological systems. We cannot expect *Phaseolus* and *Coleus* to be exceptions, since we know about the multiplicity of animal light receptors connected with circadian rhythms (p. 98) and with photoperiodism (Chapter 13).

The relations between effects of light breaks and LL are not quite as simple as might be concluded from the preceding discussion. Complications due to the occurrence of "light on" and "light off" effects in light breaks (ENGELMANN) and other details will not be discussed here. The occurrence of such effects becomes evident from the example in Fig. 76.

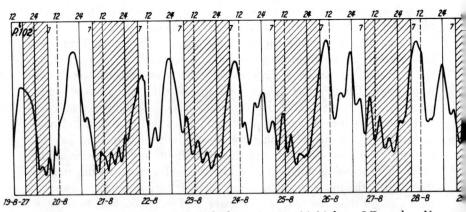

Fig. 76. *Canavalia ensiformis*, leaf movements. 24:24 hour LD cycles. No synchronization, but a rather complicated cooperation of reoccurring phase setting "light on" and "light off" effects. Dark periods shaded. After KLEINHOONTE

Light Effects

Limits of entrainment. The free-running periods of about 24 hours are normally "entrained" to exactly 24 hours by 24-hour LD cycles. Plants and animals, however, are limited in their ability to adapt their rhythms to abnormal LD cycles (*ENRIGHT; Figs. 76 to 81). The limits of entrainment depend upon the special experimental conditions and upon the species. The oscillator can be entrained not only to a 24-hour rhythm, but also in most cases to periods of 22 or 20 hours (for example, in an LD 11:11 or 10:10). But if the deviations are very large, certainly if one tries to entrain by periods of less than 16 hours, the control proves to be difficult or impossible. The organism then shows its free-running period of about 24 hours. In mice and hamsters this limit may be reached even in cycles of 21 or 26 hours (Fig. 83; TRIBUKAIT). In other animals (LAMPRECHT u. WEBER), and usually also in plants, the entrainment is still possible down to a period of 18 hours, though great irregularities already occur. Usually the upper limit of plants, animals,

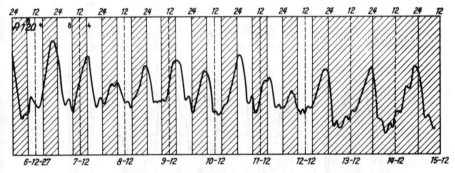

Fig. 77. *Canavalia ensiformis.* 8:8 hour LD cycles. Note synchronization, but with abnormal phases: peaks at the end of the dark period, normally (Fig. 5) at the beginning of the dark period. During the last days, DD. After KLEINHOONTE

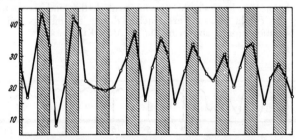

Fig. 78. *Ligia baudiniana.* Periodic variation in pigment dispersion, shown by groups of isopods kept in the laboratory. Alternating areas of gray and white correspond to LD 10:8. For details, compare Fig. 6. Maximum coloring, normally in the light period, is shifted to the dark period. After KLEITMAN 1940

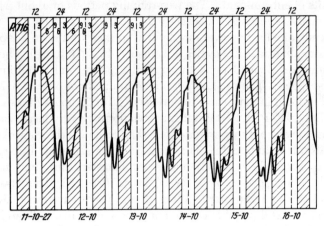

Fig. 79. *Canavalia ensiformis.* LD 6:6 hours. The physiological rhythm now continues with 24-hour periods. After KLEINHOONTE

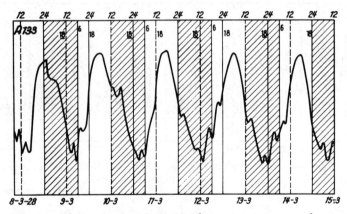

Fig. 80. *Canavalia ensiformis.* LD 18:18. The movements are synchronized, but the phase relationships between the physiological rhythm and the LD cycles are abnormal: peaks (night position) are reached during the light periods. Compare with normal relationships as shown in Fig. 5. After KLEINHOONTE

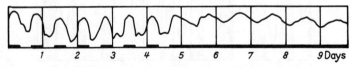

Fig. 81. *Euglena gracilis* (an algal flagellate). Rhythm in phototactic responsiveness. Entrainment to 16-hour periods by LD 8:8 hours. No continuation of these periods in DD. Broad black lines indicate darkness. Guide lines 24 hours apart. After SCHNABEL

Light Effects

and humans is about 28 to 30 hours (KLEITMAN 1939, 1949). In all cases, the limits of entrainment will be reached earlier when working with rather low intensities of light (WILKINS 1962).

Of course, a physiological process that is under control of the clock may simultaneously be under direct control of a LD cycle. Thus, a cycle deviating very much from the 24-hour cycle may enforce a corresponding physiological rhythmicity. The example reproduced in Fig. 82 clearly demonstrates that this is not an entrainment of the circadian rhythmicity. The circadian rhythm continues despite the superimposed exogenous rhythm.

> All these data refer to laboratory conditions, where the change from light to darkness and vice versa occurs abruptly by switching the illumination on or off. The reactions may be different under a gradual change, i.e., when twilight is taken into account (LEE KAVANAU).
>
> Whether or not a 24-hour LD cycle is able to synchronize the rhythm also depends on the length of the dark period. Whereas the light period may be very short (down to a few seconds or less), the dark periods of the LD cycles often must be at least 3 to 8 hours (ROBERTS 1959).

Entrainment is also possible by cycles of high and low intensity of light, instead of LD cycles. This is of some ecological interest, because of the fact that arctic plants and animals are exposed to such a condition during the season of the midnight sun. In several species of lower animals, synchronization is possible when the lower intensity is about 10% of the higher intensity. But this only seems to be true for rather low intensities (see MORI; BROWN et al. 1954). In *Phaseolus*, greater differences in the light intensities are necessary to allow synchronization (BÜNNING et al. 1966). In arctic regions up to latitudes of about 70°, the diurnal changes in the light intensities during the season of the midnight sun coupled with the simultaneous action of the temperature cycles are sufficient to entrain the circadian rhythms (REMMERT; BÜNNING et al. 1966; SWADE and PITTENDRIGH). In addition, diurnal changes

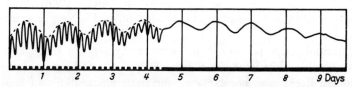

Fig. 82. *Euglena gracilis*. Rhythm in mobility. Superposition of an endogenous circadian and an exogenous 4-hour rhythm in 2:2 hour LD cycles. Circadian rhythm in DD. Broad black lines indicate darkness. Guide lines 24 hours apart. After SCHNABEL

in light quality, depending on the sun's altitude, might have an entraining effect (DEMMELMEYER u. HAARHAUS).

However, certain species of animals may shift to a free-running periodicity under the conditions of the arctic summer (ERKINARO; MÜLLER; SWADE and PITTENDRIGH; SWADE).

Frequency demultiplication. Sometimes a LD 6:6 (also 3:3 or 2:2) regulates the rhythm in the same way as LD 12:12 ("frequency demultiplication," see PFEFFER; PITTENDRIGH 1961; ROBERTS 1962; LAMPRECHT u. WEBER). This phenomenon is not surprising considering the fact that only short intervals within the physiological 24-hour cycle are particularly sensitive to light. One or several of the "superimposed" two or more 24-hour cycles will remain relatively ineffective, and only one will bring about the exact synchronization (BÜNNING u. BLUME). Let us consider for instance the effect of an LD 3:3. One of the several light pulses within each 24-hour period will be nearest to the region of maximum responsiveness in the circadian cycle, i.e., the nearest to the peak of the phase-response curve. This light pulse, in cooperation with the respective one offered 24 hours later, will have the greatest phase-shifting effect (delay or advance), whereas the other pulses will have only a comparatively small effect. Phase shifting will continue until those most effective pulses coincide with a phase that does not respond to the light.

In other cases, LD 6:6 can result in a physiological superposition of 24-hour rhythms. The phase of some of the cells or tissues is shifted then by 12 hours in relation to others.

Disappearance of the entrained periods. It hardly needs to be mentioned that after termination of the abnormal LD cycles, or if the organisms are transferred to constant conditions, LL or DD the specific oscillation period of that particular organism reappears with periods of *about* 24 hours (Figs. 77, 81). However, an organism exposed to LD cycles deviating from exactly 24 hours may choose one of the physiological rhythms available to it and perhaps maintain this for some time in LL or DD (see PITTENDRIGH 1960). This cannot be considered an exception to the above-mentioned rule. The choice is only within the limits of the possible, as has already become evident from the spontaneous frequency changes mentioned on p. 18. New frequencies are not learned.

Abnormal phase position under extreme cycles. In LD cycles deviating considerably from the natural frequency of the organism, the physiological extreme points (e.g., the extreme positions of moving leaves or the activity phases of an animal) necessarily have an entirely different relationship to the light and dark period than that normally found. In other words, the normal phase angle differences will be dis-

Light Effects

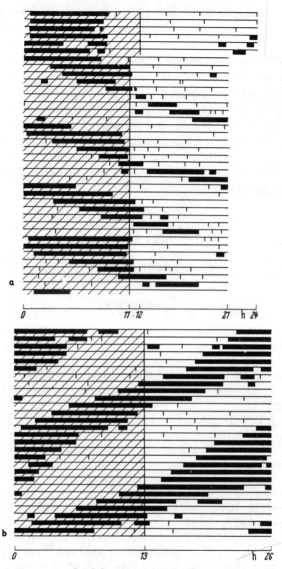

Fig. 83. Activity record of hamsters (*Mesocricetus auratus*, a night-active animal). (a) Synchronization by LD 12:12 hours (5 days). During the following 32 wrong days (LD 11:10 hours) the synchronization fails. The animal shows its individual free-running period of about 24.2 hours. During the days in which the animal's subjective night coincides with the light period, the activity is strongly suppressed by the light. Thus, the animal has a few "good nights" only in intervals of about one week. (b) In wrong days of LD 13:13 hours. The synchronization fails and the free-running periodicity becomes evident. Dark periods shaded. Light-intensity = 100 lux. Original

turbed. For example, if the dark period is too short, the organism with its own cycle length does not have enough time within the dark period to reach the usual physiological state typical of night. This state then exists only during the following light period. For the same reason, the phases usually associated with the day may occur only in the dark period following the light period. In plants, for example, which normally reach their lowest leaf position during the dark period, this state may only be attained during the light period. During the diurnal color change of the isopod *Ligia baudiniana*, the phase of most intense coloring, which usually occurs during the day, may fall into the dark period of LD 8:8 (Fig. 78; for plants, compare Figs. 77 and 80 with Fig. 5; see also Fig. 83).

References to Chapter 6
(Light Effects)

a. Reviews

*ADLER, K. 1970. The role of extraoptic photoreceptors in amphibian rhythms and orientation: a review. *J. Herpetology* **4**:99–112.—

*ASCHOFF, J. 1960. Exogenous and endogenous components in circadian rhythms. *Cold Spring Harbor Symp. quant. Biol.* **25**:11–28.—

*ASCHOFF, J., ed. 1965. *Circadian Clocks*. Amsterdam: North Holland Publishing Company.—

*ASCHOFF, J. 1965. Response Curves in Circadian Periodicity. In J. ASCHOFF, ed. Pp. 95–111.—

*ASCHOFF, J. 1969. Phasenlage der Tagesperiodik in Abhängigkeit von Jahreszeit und Breitengrad. *Oecologia* **3**:125–165.—

*ENRIGHT, J. T. 1965. Synchronization and ranges of entrainment. In *ASCHOFF (ed.), pp. 112–124.—

*HOFFMANN, K. 1965. Overt circadian frequency and circadian rule. In *ASCHOFF (ed.), pp. 87–94.—

*PITTENDRIGH, C. S. 1965. On the mechanism of the entrainment of a circadian rhythm by light cycles. In *ASCHOFF (ed.), pp. 277–297.—

*WILKINS, M. B. 1960. The effect of light upon plant rhythms. *Cold Spring Harbor Symp. quant. Biol.* **25**:115–129.—

b. Other references

ADLER, K. 1969. *Science* **164**:1290–1292.—
ASCHOFF, J. 1953. *Z. vergl. Physiol.* **35**:159–166;—
———. 1954. *Naturwiss.* **41**:49–56;—
———. 1955a. *Klin. Wschr.* **1955**:545–551;—

———. 1955b. *Naturwiss.* **42**:569–575;–
———. 1958. *Z. Tierpsychol.* **15**:1–30;–
———. 1959. *Pflügers Arch. ges. Physiol.* **262**:51–59.–
ASCHOFF, J., u. R. WEVER. 1962. *Z. vergl. Physiol.* **46**:115–128;–
———. 1963. *Z. vergl. Physiol.* **46**:321–335.
BREHM, E., u. G. HEMPEL. 1957. *Naturwiss.* **39**:265–266.–
BROWN, F. A., M. FINGERMAN, and M. N. HINES. 1954. *Biol. Bull.* **106**:308–317.–
BROWN, F. A., J. SHRINER, and C. L. RALPH. 1956. *Amer. J. Physiol.* **184**: 491–496.–
BRUCE, V. G., and D. H. MINIS. 1969. *Science* **163**:583–585.–
BRUCE, V. G., and C. S. PITTENDRIGH. 1956. *Proc. Nat. Acad. Sci. USA* **42**:676–682.–
BRUCE, V. G., F. WEIGHT, and C. S. PITTENDRIGH. 1960. *Science* **131**:728–730.–
BRUNT, E. E. V., M. D. SHEPHERD, J. R. WALL, W. F. GANONG, and M. T. CLEGG. 1964. *Ann. N. Y. Acad. Sci.* **117**:217–227.–
BÜNNING, E, u. J. BLUME. 1963. *Z. Bot.* **51**:52–60.–
BÜNNING, E., G. HAILER, u. W. MAYER. 1966. *Ber. dtsch. Bot. Ges.* **79**:7–14.–
BÜNNING, E., u. I. MOSER. 1966. *Planta (Berl.)* **69**:101–110.–
BURTON, A. C. 1956. *Canad. med. Assoc. J.* **75**:715.
CAIN, J. R., and W. O. WILSON. 1972. *J. interdiscipl. Cycle Res.* **3**:77–85.
CHANDRASHEKARAN, M. K., and W. LOHER. 1969. *J. exp. Zool.* **172**:147–152.–
CLOUDSLEY-THOMPSON, J. L. 1953. *Ann. Mag. Natur. History, Ser.* **12**:705–712.
DECOURSEY, P. J. 1961. *Z. vergl. Physiol.* **44**:331–354.–
DEMMELMEYER, H., u. D. HAARHAUS. 1972. *J. comp. Physiol.* **78**:25–29.–
DUMORTIER, B. 1972. *J. comp. Physiol.* **77**:80–112.
EHRET, CH. 1953. *Physiol. Zool.* **26**:274–300;–
———. 1960. *Cold Spring Harbor Symp. quant. Biol.* **25**:149–158.–
ENGELMANN, W. 1966. *Experientia* **22**:606–608.–
ENGELMANN, W., u. W. HONEGGER. 1966. *Naturwiss.* **53**:588.–
ERIKSON, L.-O. 1972. *Naturwiss.* **95**:219–220.–
ERKERT, H. G. 1969. *Z. vergl. Physiol.* **64**:37–70.–
ERKINARO, E. 1969. *Z. vergl. Physiol.* **64**:407–410.
FLÜGEL, A. 1949. *Planta (Berl.)* **37**:337–375.–
FRANK, K. D., and W. F. ZIMMERMAN. 1969. *Science* **163**:688–689.
GEISLER, M. 1961. *Z. Tierpsychol.* **18**:389–420.
HALABAN, R. 1969. *Plant Physiol.* **44**:973–977.–
HALBERG, F. 1959. *Z. Vitamin-, Hormon-, u. Fermentforsch.* **10**:225–296;–
———. 1960. *Cold Spring Harbor Symp. quant. Biol.* **25**:289–310.–
HALBERG, F., P. G. ALBRECHT, and C. P. BARNUM. 1960. *Amer. J. Physiol.* **199**:400–402.–
HALBERG, F., J. J. BITTNER, and D. SMITH. 1957. *Z. Vitamin-, Hormon-, u. Fermentforsch.* **9**:69–73.–

HALBERG, F., E. HALBERG, C. G. BARNUM, and J. J. BITTNER. 1959. In *Photoperiodism and Related Phenomena in Plants and Animals.* R. B. WITHROW, ed. Washington: Am. Ass. Adv. Sci.–

HASSBARGEN, H. 1960. *Z. Bot.* 48:1–31.–

HASTINGS, J. M., and B. W. SWEENEY. 1958. *Biol. Bull.* 115:440–458.–

HAUTY, G. T., and T. ADAMS. 1965. In °ASCHOFF. (ed.) 413–425.–

HEMMINGSEN, A. M., and N. B. KRARUP. 1937. *Kgl. Dansk Vidensk. Selskab. Biol. Mdd.* 13(7), 1–61.–

HILL, J. 1757. *The Sleep of Plants.* London.–

HOLDSWORTH, M. B. 1960. *J. exp. Bot.* 11:40–44.

JOHNSON, M. 1939. *J. exp. Zool.* 82:315–318.

KARVE, A. D., and S. G. JIGAJINNI. 1966. *Naturwiss.* 53:181.–

KLEIN, K. E., H. BRÜNER, H. HOLTMANN, H. REHME, J. STOLZE, W. D. STEINHOFF, and H. WEGMANN. 1970. *Aerospace Med.* 41:125–132.

KLEINHOONTE, A. 1929. *Arch. néerl. Sci. ex. et nat. IIIb* 5:1–100.–

KLEITMAN, N. 1939. *Sleep and Wakefulness.* Univ. of Chicago Press;–

———. 1940. *Biol. Bull.* 78:403–411;–

———. 1949. *Physiol. Rev.* 29,1–30.

LAFONTAINE, E., J. LAVERNHE, J. COURILLON, M. MEDVEDEFF, and J. GHATA. 1967. *Aerospace Med.* 39,944–947.–

LAMPRECHT, G., u. F. WEBER. 1971. *Z. vergl. Physiol.* 72:226–259.–

LEE KAVANAU, J. 1962. *Nature (Lond.)* 194:1293–1295.–

LOHER, W. 1972. *J. comp. Physiol.* 79:173–190.—

LOHER, W., and M. K. CHANDRASHEKARAN. 1970. *J. Insect Physiol.* 16:1677–1688.–

LÖRCHER, L. 1958. *Z. Bot.* 46:209–242.

MARTINEZ, J. L. 1972. *J. interdiscipl. Cycle Res.* 3:47–59.–

MAYER, W. 1966. *Planta (Berl.)* 70:237–256.–

MCMILLAN, J. P. 1972. *J. comp. Physiol.* 79:105–112.–

MENAKER, M. 1968. *Proc. Nat. Acad. Sci. USA* 59:414–421.–

MEYER-LOHMANN, J. 1955. *Pflügers Arch. ges. Physiol.* 260:292–305.—

MORI, S. 1944. *Zool. Mag. Tokyo* 56:1.–

MOSER, I. 1962. *Planta (Berl.)* 58:199–219.–

MÜLLER, K. 1968. *Naturwiss.* 55:140.

NOWOSIELSKI, W. v., and R. L. PATTON. 1963. *J. Insect. Physiol.* 9:401–410.

PAGE, T. L., and J. L. LARIMER. 1972. *J. comp. Physiol.* 78:107–120.–

PFEFFER, W. 1915. *Abh. kgl. Sächs. Ges. Wiss. Math.-physik. Kl.* 34:1–154.–

PITTENDRIGH, C. S. 1960. *Cold Spring Harbor Symp. quant. Biol.* 25:159–184;–

———. *Harvey Lect. Ser.* 56:93–125.–

PITTENDRIGH, C. S., and V. G. BRUCE. 1957. In *Rhythmic and Synthetic Processes in Growth.* D. RUDNICK, ed. Pp. 75–109. Princeton: Univ. Press;–

———. 1959. In *Photoperiodism and Related Phenomena in Plants and Animals.* R. B. WITHROW, ed. Pp. 475–505. Washington: Am. Ass. Adv. Sci.–

PITTENDRIGH, C. S., V. G. BRUCE, and P. KAUS 1958. *Proc. Nat. Acad. Sci. USA* **44**:965–973.–
POHL, H. 1972. *J. comp. Physiol.* **78**:60–74.
REMMERT, H. 1965. *Morph. Ökol. Tiere* **55**:142–160.–
RENSING, L., u. W. BRUNKEN. 1967. *Biol. Zentralbl.* **86**:545–565.–
ROBERTS, S. K. 1959. Ph.D. thesis. Princeton;–
———. 1962. *J. cell. comp. Physiol.* **59**:175–186;–
———. 1965. *Science* **148**:958–959.
SARGENT, M. L., and W. R. BRIGGS. 1967. *Plant Physiol.* **42**:1504–1510.–
SCHMIDT-KOENIG, K. 1958. *Z. Tierpsych.* **15**:301–331.–
SCHNABEL, G. 1968. *Planta* **81**:49–63.–
SHARP, G. W. G. 1961. *Nature (Lond.)* **190**:146–148.–
SWADE, R. H. 1969. *J. Theor. Biol.* **24**:227–239.–
SWADE, R. H., and C. S. PITTENDRIGH. 1967. *Amer. Naturalist* **101**:431–464.–
SWEENEY, B. 1963. *Plant Physiol.* **38**:704–708.
TAYLOR, D. H., and D. E. FERGUSON. 1970. *Science* **168**:390–392.–
TONGIORGI, P. 1959. *Arch. ital. Biol.* **97**:251–265.–
TRIBUKAIT, B. 1956. *Z. vergl. Physiol.* **38**:479–490.–
TWEEDY, D. G., and W. P. STEPHEN. 1970. *Experientia* **26**:377–379.–
WEVER, R. 1967. *Z. vergl. Physiol.* **55**:255–277.–
WILKINS, M. B. 1960. *J. exp. Bot.* **11**:269–288;–
———. 1962. *Plant Physiol.* **37**:735–741.–
WINFREE, A. T. 1972. *J. comp. Physiol.* **77**:418–434.–
WOBUS, U. 1966. *Z. vergl. Physiol.* **52**:276–289.
ZIMMER, R. 1962. *Planta (Berl.)* **58**:283–300.–
ZIMMERMAN, W., and D. IVES. 1971. In *Biochronometry*. M. MENAKER, ed. Pp. 381–391. Washington, D.C.: Nat. Acad. Sci.–
ZWEIG, M., S. H. SNYDER, and J. AXELROD. 1966. *Proc. Nat. Acad. Sci. USA* **56**:515–520.

7. Attempts toward a Kinetic Analysis: Models

So hört die Bewegung nicht auf, wann das Blättchen die Mitte seines Weges zurückgelegt hat, wo beyde Gegenursachen etwa ins Gleichgewicht kommen, sondern geht noch in der angefangenen Richtung fort, bis die Gegenkraft, die immer wächst, während die andere abnimmt, völlig gesieget hat.

F. v. P. SCHRANK, Vom Pflanzenschlafe und von anverwandten Erscheinungen bey Pflanzen. Ingolstadt 1792.

(Periodic movements result from overshoots and feedbacks.)

a. General remarks

Although models and mathematical treatments are very important in biological research, the history of physiology shows that in most fields of biology our models and our mathematical approaches only roughly approximated the complexity of the situation. This must be kept in mind while considering attempts to interpret the kinetics of circadian rhythms.

These remarks are not intended to deny the importance of the quantitative approach. Everybody interested in research work on circadian rhythms should familiarize himself with these analyses (*KLOTTER; *MERCER; *WEVER; *SOLLBERGER). The analyses are based on the hypothesis that the circadian rhythmicity follows simple mathematical laws (WEVER 1964, 1965). This hypothesis includes the assumption that the biological oscillation is a simple mathematical phenomenon.

At the present state of our knowledge, I prefer the opposite hypothesis, assuming very complex phenomena, for the following reasons:

1. We never observe the "basic oscillator." We only record oscillations of peripheral processes, being under control of the hidden "central clock." The chain of physiological events from the molecular basis of the clock to the overt rhythms is so complicated that the features of the overt rhythms do not necessarily reveal features of the "central clock." This becomes clear from the fact that the several overt oscillations, though apparently driven by the same basic oscillator,

may appear with different shapes; one overt oscillation being harmonic, another one being asymmetric.

2. The other complication relates to the influence of external factors. Influences of light intensity in LL (see "circadian rule") and influences of light perturbations (see "response curves") are of some importance in those mathematical approaches. These effects may vary with different mediators (central nervous system, hormones, etc.). Furthermore, these mediating processes may vary with the light absorbing organ (see Chapter 6). With the possible exception of ultraviolet light, there are no indications that the light acts directly on the oscillator.

To strengthen these general criticisms we shall see that the circadian oscillations, indeed, do not quite fit into simple mathematical models.

b. Linear and nonlinear oscillations

Are the circadian rhythms linear oscillations (harmonic, sinusoidal, pendulum-type)? Or are they nonlinear oscillations (relaxation-oscillation type)? WEVER (*1962) has discussed both possibilities. His ideas offer a good starting point for further attempts to analyze this subject. An excerpt from his arguments, in translation, follows:

> Pendulum and relaxation oscillations are different in several respects if they are considered as the extremes of a sequence with gradual transitions. The frequency of the pendulum oscillation is insensitive to external interference, but its amplitude can be influenced quite easily. On the other hand the amplitude of the relaxation oscillation is very stable, whereas its frequency can be varied easily. Another difference is the course of the oscillation: the harmonic oscillation is represented by a sinusoidal curve, the relaxation oscillation takes a more angular course. A relaxation oscillation can therefore be synchronized more easily and for a wider frequency range than the pendulum oscillation. Pendulum oscillations need many periods before they reach a new equilibrium after a change; these "adjustment processes" or "transients" affect not only the frequency but even more so the amplitude of the oscillation. For example, starting the oscillator from its state of rest results in a slow increase of the amplitude from the zero position. Only after several periods will the amplitude reach its ultimate value, which depends on the oscillation level. In contrast to this, the relaxation oscillation already reaches a state of equilibrium within the first period after a change of the external conditions. The amplitude of a relaxation oscillation in particular attains its ultimate value as early as within the first period after switching on the oscillation. Thus the amplitude of a relaxation oscillation follows an "all-or-none" rule. It can attain only one value besides zero which is given by the properties of the oscillating system,

whether the outside conditions are constant, or change periodically, or change only once, suddenly.

According to WEVER, the experimentally evaluated properties of the endogenous diurnal rhythm indicate that they are between the two types, probably resembling more the pendulum oscillation.

c. Effects of reduced energy supply

Experiments cited in Chapter 5 clearly indicate that the oscillator is not simply stopped by temperatures lower than the minimum for normal oscillation. We saw that there are phases during which low temperature permits the oscillation to continue for a few hours and other phases that do not permit this. The difference becomes noticeable in the delay of the clock, which is sometimes more and sometimes less than the period of low temperature. This will depend upon the phase of the clock at the time of treatment.

From these results we can conclude that the oscillation includes tension and relaxation processes. A chilling period of longer duration results in a gradual relaxation, approaching a certain ultimate value that does not depend on the phase when the chilling started. If the temperature subsequently rises again, the moment of the transition to normal temperatures functions as a cue for the new oscillation (see Chapter 5(d) and Fig. 58).

A relaxation oscillation can generally be expected to have two parts of unequal length, i.e., an asymmetrical course of oscillation. Extensive experiments are necessary to determine this course. Experiments with *Phaseolus* (Fig. 84) show that there are several hours within the cycle during which the clock also continues operating at low temperature, although in a reduced fashion. During the following hours, the clock is stopped nearly completely by low temperature. Here the phase shift, which is later noticeable, approximately coincides with the chilling period or is somewhat shorter than the chilling period. Only then does the phase follow during which the delay by cooling exceeds the duration of the chilling period. Apparently, the oscillator drops back during the chilling period to a lower level of tension. In other words, after restoring normal temperatures the tension has to be rebuilt from some position usually attained before the beginning of the cold treatment.

A longer chilling period always results in a final condition independent of the phase prevailing at the beginning of the low temperature (see Chapter 5(d) and Fig. 57). Evidently, the oscillator reaches a resting position. The level of this position depends, as was mentioned

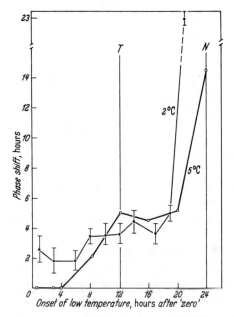

Fig. 84. *Phaseolus coccineus*, leaf movements in LL. Phase shift (delay of the clock) by 5-hour periods of 5°C and by 4-hour periods of 2°C. The chilling period started at different times with relation to the phase 0 (compare with Figs. 56 and 57). T and N indicates maxima of presumptive day and night position, respectively. After BÜNNING et al.

in Chapter 5(d), on the degree to which the energy supply is reduced. These experiments not only show quite clearly that individual phases react differently when the energy supply is turned off by low temperatures, but they also show the asymmetry characteristic of relaxation oscillations.

The special features of this asymmetry do not agree with those expected from well-known physical models of relaxation oscillations. A comparison of the behavior at 2°C with that at 5°C (Fig. 84) shows that no phases are absolutely without any energy requirement. Both curves have a stepwise character indicating a rather complex cooperation of several processes with different energy requirements.

d. Harmonic and asymmetric course

Some of the curves shown in previous chapters have a strong resemblance to harmonic (sinusoidal) oscillations (e.g., Figs. 9 and 10). Yet in some other cases curves representing physiological processes

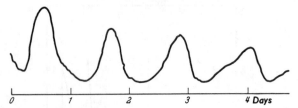

Fig. 85. *Phaseolus coccineus*, leaf movements in LL. Note very different position of minimum (physiological daytime phase) within the 4 cycles. Only the distances of the physiological nighttime peaks are constant. Original

controlled by the clock show a marked asymmetry, characteristic of relaxation oscillations.

This asymmetry may also become evident in reactions to external factors, which do not influence the oscillator as strongly as reduced energy supply. The intensively studied leaf movements of *Phaseolus* are a good example. Testing the responses of individual phases to temperature and light signals demonstrates an asymmetrical distribution of responsiveness (Fig. 50 and 70). The effects of light signals on the emergence periodicity of *Drosophila* show a similar asymmetry (PITTENDRIGH).

The nonharmonic nature of the oscillation also becomes evident in another respect. Quite often, again in the leaf movements of *Phaseolus*, only one of the two extremes (peak and trough) of the circadian cycle is developed sharply (if the rhythm is controlled by a LD cycle, it is usually the extreme in the middle of the dark period). The other one can shift considerably, even from day to day (Fig. 85). Curves of this kind strongly indicate the superposition of two or more processes.

e. Conclusions from phenomena such as fade-out and reinitiation

Damping. The possibility of damping seems to be compelling evidence for an oscillation with properties of the pendulum type; this is because relaxation oscillations follow the "all-or-none" rule. But two points have been overlooked: first, one should not draw conclusions about damping of the controlling oscillation itself from damping of a secondary physiological process *controlled* by the oscillation. For example, the activity of an animal or the growth rate of a plant can continually decrease, so that the amplitude of the rhythm of these processes becomes gradually smaller and may even fade out entirely, but the clock itself continues without being damped, as may be shown

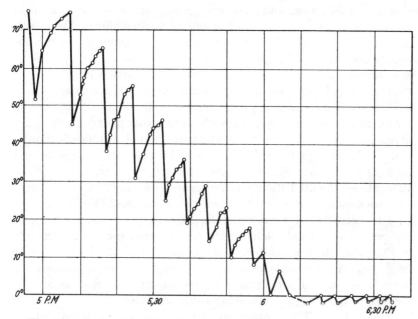

Fig. 86. *Averrhoa bilimbi*. Autonomous movements of the leaflets with strong damping at the beginning of the night. After DARWIN

by other processes also controlled by the clock. Second, the "master clock" itself, and not only the secondary oscillations controlled by it, can also show signs of damping. Although the oscillations are based on all-or-none reactions, the very same oscillator can show different amplitudes, depending on general conditions.

Biology supplies a good example of this: responses of nerve cells and similar responses of many other cells that strictly follow the all-or-none type, having an absolute refractory period, can result in responses of varying strength, depending on the physiological state. The decreased response due to a stimulus within the relative refractory period, which always follows the absolute refractory period, is the most familiar instance in which the amplitude depends on the physiological state. The "all" of a biological all-or-none reaction means the maximum possible at that particular moment. An example of damping in spite of all-or-none reactions is given in Fig. 86 (see also BÜNNING 1959; UMRATH 1959).

Model for rhythms in all-or-none systems. Strongly asymmetrical short-period oscillations in objects with all-or-none responsiveness originate from continuous stimulation or from a physiological condition ("continuous internal stimulation") leading spontaneously to repeated responses. These asymmetrical oscillations consist of a "discharge," resulting in an action, and a recovery. A new action occurs as soon as recovery reaches a certain level of the original condition. Frequency and amplitude of these spontaneous oscillations depend on the physiological condition, although they are only a sequence of all-or-none responses (examples: Figs. 86 and 89). A new action is possible, at the earliest, only after the absolute refractory period terminates. How far the *relative* refractory period has to wear off, i.e., how far the threshold that is decreasing gradually during that period has to drop, depends on the intensity of the continuous external or internal stimulation. If the change from darkness to light, i.e., establishing normal daylight conditions, suffices to induce the necessary state for "internal stimulation," the oscillations can just as well be considered as endogenous rhythmical movements that fade out in darkness (Fig. 86).

The reactions just discussed are analogous to phenomena mentioned earlier. In the range of relatively low temperatures the leaf movements of *Phaseolus* show oscillations with periods of about 7 to 14 hours (Fig. 59). A reasonable explanation is that only partial tension is possible because of a decreased energy supply, and that relaxation, therefore, begins at lower tension values than normal.

Reinitiation. The assumption of circadian relaxation oscillations is in good agreement with the reactions mentioned earlier which follow the reinitiation of a rhythm that had faded out under continuous light. We saw that a releasing dark period must last for a minimum

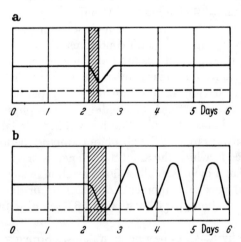

Fig. 87. Diagram to explain the initiation of a rhythm by single dark periods offered during LL. (a) the dark period is not long enough to force the tension of the oscillator down to a critical value (broken line). (b) the longer dark period forces the oscillator down to the critical level and allows the initiation of the oscillation.

length of time. Apparently, the oscillator must be "lowered" to a level that gives the impulse for further oscillation. Shorter dark periods have no effect (see Figs. 31 and 87). These reactions could not be explained as satisfactorily on the basis of a simple pendulum oscillation.

f. Effect of synchronizers on a relaxation oscillation

Both circadian and relaxation oscillations can be synchronized easily and within a rather wide range of entrainment by external oscillations. The synchronizing functions can be illustrated by making the not improbable assumption that light or an increase in temperature promotes the tension process. In other words, the tension can rise slightly beyond the value usually possible, before the relaxation process begins. Light or a rise in temperature coinciding with the relaxation process may interrupt this by the premature initiation of new tension (Fig. 88).

The oscillator is controlled by the 24 hours LD cycle so that the transition from tension to relaxation coincides approximately with the middle of the dark period, if the available evidence permits such a generalization.

g. The transients and the response curves

The occurrence of transients was used as an argument in favor of harmonic (pendulum) oscillations (see page 117). Cycles influ-

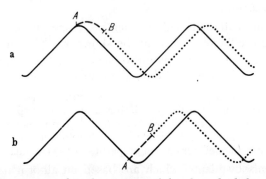

Fig. 88. Diagram to explain the resetting of the phases by light or temperature signals. (a) light offered from A to B, i.e., during the end of the tension process allows continued tension to a stronger degree. (b) light offered from A to B, i.e., now during the end of the relaxation, causes earlier occurrence of the new tension.

enced directly by light or temperature deviate from the usual length. This can be explained by the model of a relaxation oscillator, as we saw in the aforementioned examples. A different explanation seems to be necessary for transients, since they extend over several cycles.

Coupled oscillators. BRUCE and PITTENDRIGH (see *PITTENDRIGH; *PITTENDRIGH and BRUCE) have suggested a model of coupled oscillators to explain the transients. There are several facts in favor of such a model. For example, one might imagine that the phases of one participating oscillation are determined immediately and permanently by a temperature or light signal in the way already described, whereas the phases of a second oscillator linked to the first one would be determined only after several cycles.

This explanation may apply in certain cases, especially in higher animals. For higher animals we know the importance of different time requirements for synchronizing the several rhythms and the role of mutual entrainment (see Chapters 6 and 7).

Yet investigations on the diurnal petal movements of *Kalanchoe blossfeldiana* do not support such an interpretation in this case. Transients can be induced by short light signals during DD (Fig. 75). The different phases of the transient cycles show the same reactions to light interruptions as their counterphases in the normal petal movement cycle. The overt transient cycles therefore reflect the behavior of the underlying oscillator. They cannot be interpreted as the resultant between one oscillator that is immediately reset by the first light signal and another oscillator that does not follow immediately.

> Moreover, it is very difficult to understand "reversing transients" on the basis of coupled oscillators. (In reversing transients, the first shift has a direction different from those that ensue; e.g., the first transient cycle is shorter, the ensuing one longer than those of the steady state.)

Relaxation oscillations with transients. Transients are also possible in relaxation oscillations. This can be shown easily with a biological model. We shall again use the movements of leaves with periods of a few minutes, which are based on all-or-none responses.

> We could also refer here to the responses in nerves and muscles. The physiological processes during all-or-none responses in plants appear to be related to those in animals, but in plants these processes take place much more slowly. It is therefore an advantage to use them as a "model" for certain properties of the circadian rhythm.

Movements by plants, which are based on all-or-none responses, were thoroughly investigated decades ago. We know that processes with several symptoms of nervous actions are responsible for these movements. Action currents as well as an absolute refractory period

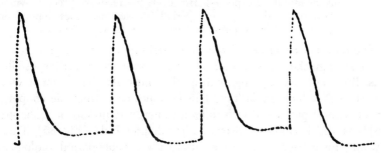

Fig. 89. *Mimosa pudica*. Periodic reaction due to continuous electric stimulation. The time between the individual reactions is several minutes. After BOSE, from BÜNNING 1959

followed by a relative refractory period can be observed. These movements can be released by touching (thigmonasty, seismonasty) and, in some cases, also by light stimulation.

It has already been mentioned that under favorable light and temperature conditions movements such as these may appear as "autonomous" movements in some objects. This means that an endogenous rhythm is observed. All transitions exist between these types. For example, we can also cause periodical movements by continued stimulation in objects that usually show only a single response (Fig. 89). But this is also sometimes possible with a single strong stimulus. This strong stimulus releases so much excitation substance (a substance actually found in extracts) that enough of that substance remains for a second and third response after the first one has worn off. This means that the wearing off of the refractory periods is not identical with the destruction of the excitation substance. (The "action substances" in nervous actions may also be detectable longer than the response.)

Processes such as these show that the total response results from a disintegration and a recovery process; recovery begins before disintegration has ended. An early release of the recovery process keeps the disintegration process and the resulting reaction from reaching a maximum value which could be possible without recovery. The recovery process also causes the refractory periods to wear off gradually. The recovery process corresponds with the energy-consuming tension process of a relaxation oscillation in contrast to the relaxation process.

> It has long been known that the energy relations are similar to this during responses of the described type. The disintegration processes are also possible under oxygen deficiency and under the influence of respiration inhibitors. They can even be released by these

agents, as might be expected. However, the recovery process, which is necessary for the refractory periods to wear off, does depend on energy supplied by respiration (see GUTTENBERG u. REIFF).

The fact that processes of recovery begin before disintegration has ended distinguishes these oscillations from ordinary relaxation oscillations. But some overlapping such as this must also be assumed with respect to circadian oscillations. This becomes clear from the cooperation of several processes with different energy requirement (Chapter 8(c)), as well as from the discussion in Chapter 7(d) (Fig. 85). Thus, it would be inappropriate to compare these short-period biological oscillations as well as the circadian ones with the usual textbook models for relaxation oscillations.

If we cause a periodical action of the specified type (Fig. 89) by continuous stimulation, we can disturb this rhythm by a strong single extra stimulus in the following way (Fig. 90a): the stimulus causes a new response before the relative refractory period has worn off to the level that would have permitted another response by the continuous weak stimulation. This reaction occurring in an early part of the relative refractory period is smaller than normal, which is characteristic

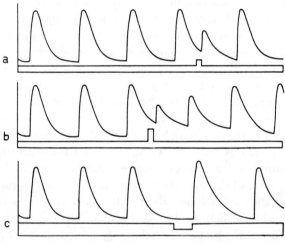

Fig. 90. Oscillations that result from responses of the all-or-none type. Horizontal line above abscissa indicates continuous stimulation. (a) transitory increase of stimulation-intensity during the indicated time. The relative refractory period is interrupted earlier than before. The refractory period of this premature excitation is shorter than the normal one. (b) very strong transitory increase of stimulation. Due to accumulation of excitation substance premature interruption of refractory periods occurs throughout several cycles. (c) transitory decrease of stimulation-intensity at indicated time. The relative refractory period is interrupted later than before. Therefore, the next response is stronger and has a longer refractory period.

Attempts toward a Kinetic Analysis: Models

of responses by a strong stimulus within the relative refractory period. Since this response is smaller, its refractory period *can* also be correspondingly shorter. This was observed in experiments with both animals and plants (see, e.g., UMRATH 1928; BÜNNING 1929). Consequently, the subsequent reaction will inevitably occur somewhat earlier, although no other strong single stimulus had been offered. Even later "cycles" of this rhythm may show "transients," if the single stimulus was strong enough to cause an extreme accumulation of excitation substance, as discussed on page 125. Under these circumstances, the refractory period can be broken prematurely during several periods (Fig. 90b).

In such a system we can also cause transients consisting of lengthened periods, by decreasing temporarily the strength of the continuous stimulation. Now the refractory period can wear off later than usual. This makes the following response stronger, which, in turn, inevitably results in a longer refractory period than the preceding one. As a result, the third reaction is delayed (Fig. 90c).

In the autonomous, short-period movements of the leaflets from *Desmodium gyrans*, single external stimuli can, in fact, cause transients in these simple relaxation oscillations as with the diurnal oscillations. These movements appear to provide a good example for this "biological

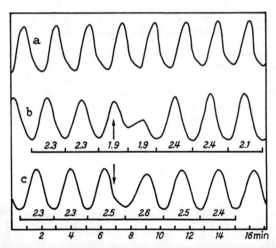

Fig. 91. *Desmodium gyrans*. Autonomous movements of the leaflets with periods of a few minutes. (a) undisturbed movement (continuous light, constant temperature of 27°C). (b) a shock of low temperature (0°C, 15 sec) was offered at the phase indicated by the arrow. Note advance connected with transients. (c) the shock of low temperature was offered at another phase. Note delay connected with transients. Length of periods in minutes. After CIHLAR

model" of circadian rhythms. They have some features in common with circadian rhythms: the oscillations are approximately, but not fully, harmonic; phase shifts with transients can be induced; response curves rather similar to those of circadian rhythms are possible (BÜNNING u. ZIMMER, Figs. 91, 92). Another example for phase shifts in short-period oscillations in plants is shown in Fig. 93.

Note that we are referring here in particular to movement responses and autonomous movement rhythms in plants. But response curves similar to those just mentioned are known in many other cases of biological rhythms with short periods. The role of these response types for the synchronization with external rhythms (especially, periodic electrical stimulation) is well known. For nerves, see FESSARD; ARVANITAKI. For other animal rhythms, see MARX et al. For electric rhythms in large plant cells, see ALBE.

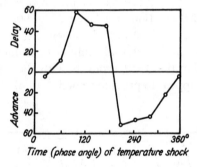

Fig. 92. *Desmodium gyrans*. Response curve resulting from a greater series of experiments such as those explained in Fig. 91. Delays and advances are expressed as percent time difference in completing three cycles after shock of low temperature. Abscissa: phases of the cycle of approximately 2 minutes. Phase 0° corresponds to maximum upward position of the leaflet, phase 180° to the maximum downward position. After CIHLAR

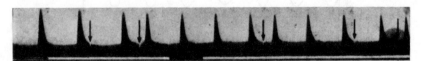

Fig. 93. Example for phase shifts (advances or delays) in short-period oscillations. Electric oscillations in the alga *Chara*, influence of electric stimulations (indicated by arrows), applied at different phases of the refractory periods. Black interruption of white line on bottom indicates 1 minute. From AUGER

h. Refractory periods in the circadian cycle

The comparison between the well-known short-period biological rhythms and the circadian rhythms should be pursued to the phenomenon of refractory periods.

In the short-period cycles of nerves, other animal or plant organs, cells, etc., each individual reaction in the oscillation is of the all-or-none type, and is therefore accompanied by an absolute and an ensuing relative refractory period. The phase-response curves relating to circadian cycles show refractory periods, too. An absolute refractory period is not detectable in a phase-response curve like that in Fig. 73, but it might be concluded from such cases as shown in Figs. 71 and 72. In these examples, there are several phases not showing any response

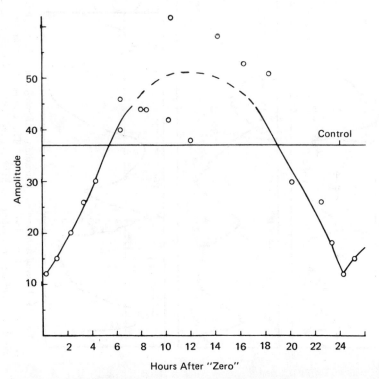

Fig. 94. Result of analyzing a greater number of experiments with *Kalanchoe*, of the type shown in Fig. 75. Advances (after light breaks from about hour 0 up to nearly hour 12) are like "extra systoles" characterized by lower peaks. Delays (after light breaks later than about hour 12) are characterized by higher peaks. After Bünning u. Blume

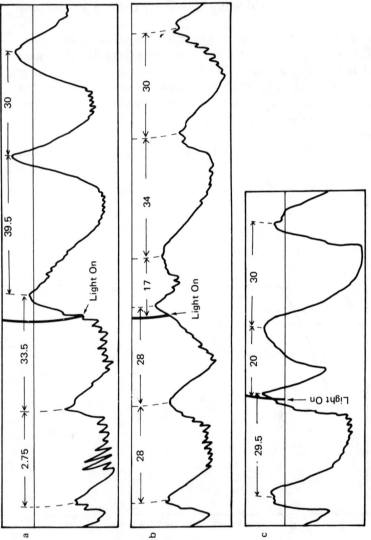

Fig. 95. *Phaseolus coccineus*. Circadian leaf movements in continuous darkness which is replaced by continuous light at the indicated times. The curves show that, depending on the circadian phase during which the shifting to continuous light started, either a lengthening of the affected cycle appears, or a new reaction is induced. The new reaction (similar to an "extra systole") overlaps the preceding one. After BÜNNING a. MOSER

Attempts toward a Kinetic Analysis: Models 131

to a light pulse. However, an increase in intensity or duration of the light pulse offered during DD causes an abbreviation of the "absolute refractory period" within the cycle (WINFREE). That means, increasing the strength of the light pulse makes cases like those of Figs. 71 and 72 shift toward the type represented in Fig. 73. Thus, the region of no response to a weak light pulse has features of a relative refractory period: the threshold values for responses are higher.

A relative refractory period is always recognizable. The amplitudes and periods of responses induced by a light pulse depend on the phase. This is already evident in the example in Fig. 75 and becomes clearer by a mathematical analysis of the respective experiments (Fig. 94).

Relative refractory periods of short-period oscillations in nerves, etc. are characterized by lower possible amplitudes and higher threshold values. This holds also for the circadian cycles, as was already

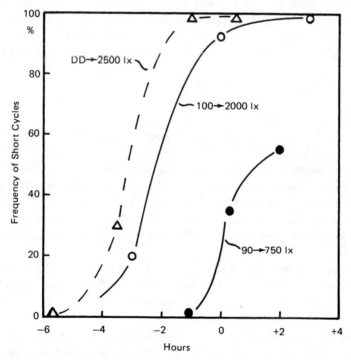

Fig. 96. Experiments such as those shown in Fig. 95. The curves show how often (in terms of percent) the replacement of continuous darkness by continuous light induced new reactions (as in Fig. 95). Abscissa: time of beginning continuous light, hours before (−) and after (+) the peak of the already occurring reaction. The data show that this new induction becomes easier the later the light is switched on, and the higher the intensity is. These are typical features of a relative refractory period. After BÜNNING u. MOSER

concluded from the abbreviation of the nonresponding period while applying stronger stimuli. Fig. 95 brings additional evidence. In these experiments with *Phaseolus*, the plants were brought from DD to LL or from LL of low intensity to LL of a higher intensity. Fig. 95 shows that this transfer can result in two responses: Either the period during which the transfer occurred is strongly increased (Fig. 95a) or a new cycle is induced, which overlaps with the one being treated (Fig. 95b, c). Fig. 96 shows that the frequency in the appearance of these new cycles depends not only on the phase during the transfer (i.e., on the wearing off of the relative refractory period), but also on the height of the step. A high step (from DD to LL 2500 lux) allows the new cycles to occur about 4 hours earlier in the relative refractory period than the rather low step from LL 90 lux to LL 750 lux.

There is some evidence that circadian rhythms in animals reveal the same features as described in Fig. 95. The first cycle after a transition from DD to LL can be much longer than the following steady state cycles. This overshoot may reach values of more than 2 hours (in the beetle *Tenebrio molitor*, LOHMANN), or sometimes even values between about 1 and 4 hours (in the cockroach *Blaberus craniifer*, WOBUS). Though this overshoot is not as high as in *Phaseolus* (Fig. 95a), it is a similar phenomenon.

i. Conclusions

This chapter has sought mainly to indicate some striking similarities of the circadian rhythms found in better-known short-period oscillations in plants and animals. It is evident that attempting to analyze circadian rhythms mathematically, on the lines of well-known oscillations in physics and technology, cannot by itself yield a satisfactory model.

Recently, JOHNSSON and KARLSSON (also KARLSSON and JOHNSSON) presented a feedback model showing many features of circadian rhythms (phase shifts, entrainment, transients, damping). Many experimental results support this model.

For the reasons cited in the general introduction to this chapter, all these attempts are of a rather restricted value. As HASTINGS and KEYNAN wrote: "For an ultimate understanding of these systems we must turn to the genetic and molecular level; inevitably we are seeking the solution to the question of mechanism in terms of chemical activities and transformations."

References to Chapter 7
(*Attempts toward a Kinetic Analysis: Models*)

a. Reviews

*KLOTTER, K. 1960. General properties of oscillating system. *Cold Spring Harbor Symp. quant. Biol.* **25**:185–187.
*MERCER, D. M. S. 1960. Analytical methods for the study of periodic phenomena obscured by random fluctuations. *Cold Spring Harbor Symp. quant. Biol.* **25**:73–86.
*PITTENDRIGH, C. S. 1960. Circadian rhythms and the circadian organization of living systems. *Cold Spring Harbor Symp. quant. Biol.* **25**:159–184.–
*PITTENDRIGH, C. S., and V. G. BRUCE. 1957. An oscillator model for biological clocks. In *Rhythmic and Synthetic Processes in Growth.* D. RUDNICK, ed. Princeton Univ. Press. Pp. 75–109;–
———. 1959. Daily rhythms as coupled oscillator systems and their relation to thermoperiodism and photoperiodism. In *Photoperiodism and Related Phenomena in Plants and Animals.* R. B. WITHROW, ed. Pp. 475–505. Washington: Am. Ass. Adv. Sci.
*SOLLBERGER, A. 1965. *Biological Rhythm Research.* Amsterdam: Elsevier.
*WEVER, R. 1962. Zum Mechanismus der biologischen 24-Stunden-Periodik. *Kybernetik* **1**:139–154;
———. 1964. Zum Mechanismus der biologischen 24-Stunden-Periodik. *Kybernetik* **2**:127–144;–
———. 1965. A mathematical model for circadian rhythms. In *Circadian Clocks.* J. ASCHOFF, ed. Amsterdam: North Holland Publishing Company. Pp. 47–63.

b. Other references

ALBE, D. 1940. *C. r. Soc. Biol.* **135**:1563–1565.–
ARVANITAKI, A. 1943. *Arch. Int. Physiol.* **53**:533–559.–
AUGER, D. 1936. *Comparaison entre la rythmicité des courants d'action cellulaires chez les végétaux et chez les animaux.* Paris. Hermann et Cie.
BÜNNING, E. 1929. *Z. Bot.* **21**:465–536;–
———. 1959. *Handb. Pflanzenphysiol.* **17**(1):184–238;–
———. 1963. *Z. Bot.* **51**:174–178.–
BÜNNING, E., u. J. BLUME. 1963. *Z. Bot.* **51**:52–60.–
BÜNNING, E., S. KURRAS, u. V. VIELHABEN. 1965. *Planta (Berl.)* **64**:291–300.–
BÜNNING, E., u. I. MOSER. 1967. *Planta (Berl.)* **77**:99–107.–

BÜNNING, E., u. R. ZIMMER. 1962. *Planta (Berl.)* **59**:1–14.
CIHLAR, J. 1966. Dissertation. Tübingen.
DARWIN, C. H., and F. 1880. *The Power of Movement in Plants.* London: J. MURRAY.
FESSARD, A. 1942. *C. r. Soc. Biol.* **136**:268–272.
GUTTENBERG, H. v., u. B. REIFF. 1958. Planta **50**:498–503.
HASTINGS, T. W., and A. KEYNAN. 1965. In *Circadian Clocks.* J. ASCHOFF, ed. Amsterdam: North Holland Publishing Company. Pp. 167–182.
JOHNSSON, A., and H. G. KARLSSON. 1972. *J. theor. Biol.* **36**:153–174.
KARLSSON, H. G., and A. JOHNSSON. 1972. *J. theor. Biol.* **36**:175–194.
LOHMANN, M. 1964. *Z. vergl. Physiol.* **49**:341–389.
MARX, CH. H., F. ISCH, et F. ROHMER. 1950. *Revue Neurologique* **82**:1–6.
PITTENDRIGH, C. S. 1966. *Z. Pflanzenphysiol.* **54**:275–307.
UMRATH, K. 1928. *Z. Biol.* **87**:85–96;–
———. 1959. *Handb. Pflanzenphysiol.* **17**(1):24–110.
WINFREE, A. T. 1970. *J. theor. Biol.* **28**:327–374.–
WOBUS, U. 1966. *Biol. Zentralbl.* **85**:305–323.

8. Attempts toward a Biochemical and Biophysical Analysis

a. Rhythms in enzyme activity

Examples of enzyme rhythms. Diurnal changes of respiration in plants and animals and of the photosynthetic capacity in plants were observed quite early in research. These changes suggested that diurnal rhythm might be explained by the accumulation of respiratory or photosynthetic products. One might imagine that an accumulation of these substances could cause an inactivation of enzymes, after which the substances would then gradually decompose. This interpretation proved to be incorrect, as have all the other similar explanations based on simple metabolic processes.

Fluctuations of enzyme activities have been observed repeatedly, not only indirectly through changed metabolic activities but also by enzyme extraction. Luciferase and ribulose diphosphate carboxylase in *Gonyaulax* (HASTINGS and SWEENEY; SWEENEY) and several other enzymes in lower and higher plants provide examples (EHRENBERG; VENTER; RICHTER u. PIRSON). Diurnal fluctuations of enzyme activity have also been evaluated directly in animal tissues; for example, in the adrenal gland (see GLICK et al.), in the liver (GLICK and COHEN; HARDELAND), and in kidneys (°VAN PILSUM and HALBERG). LE BOUTON and HANDLER showed that the circadian rhythmicity of protein synthesis in the liver of rats continues also under conditions of starvation. Thus, this rhythm is not due to the feeding rhythm.

Different factors may be involved in these circadian rhythms of enzyme activity: action of inhibitors, inactivation by steric alteration, and rate of degradation. But circadian rhythms even in the *de novo* synthesis of enzymes are known (HARDELAND, with rat liver tryptophan pyrrolase). According to experiments by SULZMAN and EDMUNDS, the circadian oscillations in alanine dehydrogenase activity in *Euglena* are due to circadian oscillations in syntheses and destruction.

Suppression of enzyme activity. Yet, none of these enzyme fluctuations have been proved to be part of the clock mechanism. They are, at least in many cases, obviously only processes that are controlled by the clock, just as fluctuations in growth rate, turgor pressure, or running activity are only controlled processes. Three facts make this conclusion inevitable. First, there is no relationship between the length of the periods or their parts and the intensity of these metabolic activities. Second, enzyme inhibitors can decrease the amplitude of the rhythms, but they do not necessarily suppress the clock. Third, supplying the substrates necessary for a certain reaction or changing the temperature results in an acceleration or inhibition of the metabolic processes, but it does not accelerate or inhibit the rhythm.

Photosynthesis in the alga *Gonyaulax* can be completely suppressed by dichlorophenyl dimethyl urea (DCMU), yet the clock continues to run: the rhythm of bioluminescence is not inhibited. The bioluminescence glow can be inhibited by puromycin, an inhibitor of protein synthesis. But the rhythm resumes without phase shift after removal of puromycin (*Hastings 1970). Feldman (1967) published some evidence for the assumption that protein synthesis itself is involved in the operation of the clock. The period of the rhythm of photoactic response in *Euglena* is increased by cycloheximide. But this was not connected with a phase shift. Feldman (1968) also reported on a circadian rhythm of amino acid incorporation in nondividing cells of *Euglena*.

With the help of such substances as NaCN, arsenate, 2,4-dinitrophenol, and NaF, Bühnemann was able to influence the rhythm

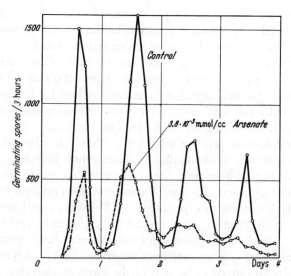

Fig. 97. Spore discharge in *Oedogonium cardiacum* in LL under the influence of Na_2HAsO_4. After Bühnemann 1955

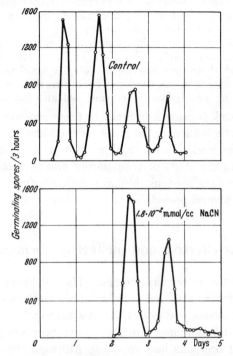

Fig. 98. Spore discharge in *Oedogonium cardiacum* in LL under the influence of NaCN. While the concentration of the poison decreases gradually, the periodicity starts again without shifting the phases. After BÜHNEMANN

of sporulation in the alga *Oedogonium*, but this only suppressed the amplitude of the overt rhythm. As long as the discharge of spores could be observed, it was cyclic without any lengthening of the periods or shift of the phases (Fig. 97). The effects of cyanide were especially interesting in these experiments. After a long application of cyanide, and as its concentration slowly decreases, the periodic liberation of spores may again become evident, but even then the phases have not been shifted. The maxima and minima still appear at the same time as in the controls (Fig. 98). Thus, the endodiurnal system continues, even when the processes controlled by it can no longer be recognized.

b. Earlier reports about chemical effects

In the older literature some findings seem to indicate a strong metabolic influence on the endogenous rhythm. Withdrawal of oxygen or addition of CO_2 was found to cause an inhibition in plants (leaf movements of beans) and in insects (time sense, emergence rhythm).

GRABENSBERGER mentions an acceleration of the clock in bees (earlier appearance at the feeding place) caused by increased metabolism. Quinine, on the other hand, was supposed to have a delaying effect (KALMUS). A narcosis with ether had no effect. Oxygen withdrawal for several hours delays the clock in *Drosophila*. RENNER (1957) has searched without success for delay or acceleration due to substances inhibiting metabolism (quinine, see also WERNER) or accelerating metabolism (thyroxine, see also RENNER 1958).

One drawback is present in many of these experiments: the factor examined directly influences the process dependent on the endodiurnal rhythm, but the endodiurnal rhythm itself may remain unchanged. Therefore, as in the study of temperature effects, a conclusion about influences on the endogenous rhythm is only possible if effects can be observed over several periods.

c. Various effective and ineffective chemical factors

Variety of the tested substances. The effectiveness of a large variety of substances has been tested in plants and animals (*HASTINGS 1960, 1970; *HASTINGS and BODE). Besides the usual enzyme inhibitors, many organic and inorganic substances have been examined in beans. Colchicine and urethane had an effect. Although they first caused a lengthening of the periods, later the periods might even be shortened (BÜNNING 1956, 1957).

Many different growth hormones, vitamins, etc., have been tested in *Oedogonium*, *Gonyaulax*, and *Phaseolus*. The possible influence of growth hormones on the growth rhythm in *Avena* coleoptiles has also been examined. BALL and DYKE could find no clear-cut effect.

Substances that are effective in beans may have little or no effect in other objects. It must remain in doubt, then, whether the effects of the compounds are specifically on the clock itself or only influences upon some processes controlled by the clock. BRUCE and PITTENDRIGH reported many fruitless efforts in *Euglena*.

Effects of extreme reduction of respiration. Obviously, the circadian clock continues undisturbed even when the turnover of energy has dropped to a small fraction of the normal. We were forced to conclude this earlier from the experiments with low temperatures. The clock stops only when respiration has dropped to a very low percentage of the intensity under optimum temperatures. We must draw the same conclusion from experiments on the influence of respiration inhibitors. On the other hand, an inhibitor restricting respiration may prevent the overt periodicity without delaying the clock itself (Fig. 98). One should not conclude from this that the operation of the clock is completely independent of respiration or that it depends only upon a certain component not being affected by the inhibitor. Yet it is safe to say that the amount of energy required for the continuation of the clock is smaller than that for carrying out the controlled processes (e.g., growth, running activity, and spore discharge). The inhibitor concentration must be close to lethal to stop the clock by reducing respiration. If this degree of inhibition is accomplished, one can in fact observe effects upon the course of the clock. But so far there have been no quantitative studies on the relationship between reduction of respiration and type or degree of influence on the clock (see the data of *HASTINGS and of KELLER).

Old autotrophic cultures of *Euglena* do not show a rhythm in respiration, and under certain conditions they show no measurable respiration at all. Nevertheless, the circadian rhythm in motility persists (BRINKMANN 1971).

The influence of oxygen withdrawal has been studied repeatedly by replacement with nitrogen. An example is given in Fig. 99. The peak following the oxygen withdrawal is delayed considerably, but the following periods are again shortened. This reaction is also characteristic of other organisms. We must ask ourselves whether the influence might not be upon a process controlled by the clock (in the example given, on growth). But lasting phase shifts also occur.

Inhibitors and oxygen withdrawal have different effects on the different phases of the oscillations, as indicated by the data of *HASTINGS and of KELLER, but far-reaching conclusions are not yet possible.

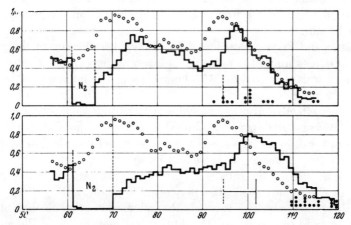

Fig. 99. Growth rhythm of *Avena* coleoptiles. Air was replaced by N_2 at indicated times. After BALL and DYKE

WILKINS and WARREN, investigating the circadian growth rhythm in *Avena* seedlings, found delays by several hours during oxygen depletion. These authors also checked the dependence of this effect on the phase during which oxygen was removed. Lack of oxygen resulted in a shift of the first peak after treatment, but not necessarily in a shift of the second peak.

Phaseolus and *Kalanchoe* were exposed to anaerobic conditions at different phases. Certain phases in the cycle did not respond with shifts, but other phases did (BÜNNING et al. 1965). Once more an asymmetry of the response is demonstrated and may be compared with the responses to low temperature (see Chapter 5(d)). A phase-depending difference of response to anaerobiosis was also observed by WILKINS.

> Poisons like DNP or KCN may also induce phase shifts; both delays and advances can result from such treatments (STEINHEIL). Thus, there is apparently a rather complex cooperation of different partial processes.

Period lengthening by chemicals. According to the observations mentioned above, a decrease in period length as a result of poisoning is not surprising. Lengthening of the period in response to the continuous influence of some substances is more remarkable. In *Phaseolus*, ethyl alcohol (2%), theobromine (0.1%), and theophylline (0.1%), offered via the transpiration stream, cause a lengthening of the periods by 2–4 hours (KELLER). This lengthening of the periods occurred not only immediately after the beginning of the chemical influence, but for at least 4 or 5 days. In other words, these are not just transient reactions. The strongest effect in *Phaseolus* was with ethyl alcohol. ENRIGHT (1971b) found that 0.5% of ethyl alcohol increases the period of the tidal rhythmicity of the isopod *Excirolana chiltoni* on the average by about 1 hour.

BRUCE and PITTENDRIGH found an obvious lengthening of the periods in *Euglena* under the influence of heavy water (D_2O). The period of the controls was 23.5 to 23.75 hours; under the influence of heavy water, the periods changed from 26.5 to 28 hours. Heavy water proved to be very effective also in *Phaseolus*. The periods may increase by 6 hours (BÜNNING u. BALTES). Adding D_2O to the drinking water of mammals can result in an increase of the period of activity rhythm by 6 or 7% (SUTER and RAWSON; PALMER and DOWSE).

At least some of the substances causing considerable period lengthening by continuous application may result in phase shiftings if applied for a short time only. This is true of the influence of ethyl alcohol, other alcohols, and heavy water on the diurnal leaf movements of *Phaseolus* (Fig. 100).

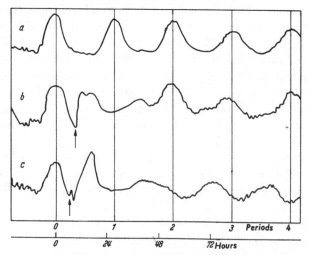

Fig. 100. *Phaseolus coccineus*, leaf movements in LL. Effect of single exposure to ethyl alcohol for 3 hours, beginning at time indicated by arrow. The alcohol was offered by putting cut plants into water with 25% alcohol. The concentration that actually reached the leaves was much less. After BÜNNING u. BALTES

d. The possible role of enzyme rhythms with shorter periods

Information on oscillations in enzyme activity is rapidly increasing. These oscillations are due to feedback loops resulting, in some cases, from effects of metabolites on particular enzymes, or in other cases from inhibition of enzyme production along the well-known chains from gene via mRNA.

In yeast cells, for example, the synthesis of one of the enzymes occurs with periods of a few hours (BERNHARDT et al.). The period of oscillation in enzyme activity may vary between a few hours to less than 1 hour. For further references, see CHANCE et al. 1964, 1965; BETZ and CHANCE; MASTERS and DONACHIE; and *GOODWIN. The nature of these oscillations is different in the several cases, both with respect to the biochemical mechanism and to the factors influencing the shape of the oscillations. In certain cases they are sinusoidal, in others strongly asymmetric.

In all cases the periods are much shorter than 24 hours. It has been suggested that circadian rhythms may result from the cooperation of shorter rhythms with different periods. This approach faces the same difficulties as an explanation on the basis of frequency demultiplication. *GOODWIN offers the following argument in favor of this hypothesis. After an artificial LD regime with cycles shorter than 24 hours, the organism, in constant conditions "will continue to show an endogenous rhythm which is the same as that into which it was forced, thus demonstrating the existence of internal oscillations with a period shorter than 24 hours." This does not, however, agree with the facts. A synchronization of the endogenous diurnal periodicity to 12-hour cycles is impossible (see Chapter 6) and, furthermore, there is no indication of 12-hour cycles after treatment with external cycles of this type. In addition, all the tests on responsiveness of the several phases in the circadian rhythm have not revealed any indication for such a synthesis by shorter cycles.

The strong temperature dependence of the periods in the short-period oscillations, showing Q_{10} values up to 4, offers another argument against their involvement in circadian rhythmicity (references: BRINKMANN 1971).

Nevertheless, we are not yet allowed to exclude the possibility that populations of interacting biochemical oscillators build up the circadian oscillation. PAVLIDIS (1969, 1971) has simulated such a system on the computer, and he found features of circadian rhythmicity, among them rhythm splitting. However, in living systems this splitting (Fig. 133) is known only for multicellular organisms. All the attempts to find something like this in unicellular organisms were in vain. In the unicellular alga *Gonyaulax,* the phase relationships among the several circadian rhythms remain unchanged, even after exposure to constant conditions for several weeks, or after a phase shift by offering a 6-hour dark period during LL (MCMURRY and HASTINGS). This is strong evidence for the presence of a single "master clock."

Interesting experiments by JACKLET and GERONIMO with isolated eyes of the sea hare *Aplysia* suggest a population of clocks. In this case, the circadian rhythm in the frequency of compound action potentials depends on the interaction among the cells of the retinal population. Reducing the population number results in shorter periods, and below a critical number oscillations with much shorter periods occur. These results still allow different explanations.

e. Biochemical oscillations with longer periods

Biochemical cycles with periods of several hours offer another approach to models for circadian rhythms. PAVLIDIS and KAUZMANN found that a rather simple biochemical oscillator can simulate the features of circadian rhythms. "The system involves two enzymes whose concentrations are assumed to be decreasing functions of temperature. This results in a temperature-independent period of oscillation. The

Attempts toward a Biochemical and Biophysical Analysis   143

feedback necessary for oscillations is achieved through an allosteric effect."

Experimental evidence for biochemical oscillations with rather long periods, though involving only a few enzymes, is offered in several papers. TSCHUDY et al. found that single intravenous injections of estradiol given to ovariectomized rats produce irregular oscillations of the levels of two enzymes. The periods were about 10 hours. KLITZING, working with the giant cells of *Acetabularia*, found oscillations in the activity of enzymes and in the concentration of ATP with periods between 2 and 360 minutes.

We cannot yet decide whether or not biochemical oscillations like these offer a satisfying model for the circadian clock.

f. Role of the nucleus

Microscopic changes in the nuclei. The diurnal fluctuations in volume are worthy of special mention (Fig. 101, Chapter 3(b) and Chapter 4(c)). Sometimes the volume more than doubles within a few hours of the day and then decreases again during the following hours (BÜNNING u. SCHÖNE-SCHNEIDERHÖHN).

The nuclei have a more compact structure when they are small than when swollen by water intake. These volume fluctuations continue under constant conditions. In mice, diurnal fluctuations in the volume

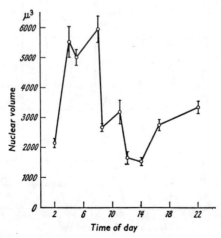

Fig. 101. Diurnal changes in the volume of nuclei (medium values) in guard cells of *Allium cepa*. Second day of LL and constant temperature. After BÜNNING u. SCHÖNE-SCHNEIDERHÖHN

of the cell nuclei have also been found—for example, in the neurosecretory system of the hypothalamus. NIEBROJ considers this as a possible cause for other diurnal body functions.

For further examples, see HOFER u. BIEBL, BIEBL u. HOFER, SCHÖLM (plant cells), MÖDLINGER-ODORFER (animal cells with neurosecretory functions), and QUAY and RENZONI (in the pineal). In most cases, not only changes in the volume but also accompanying changes in the structure of the nucleus were observed, as well as remarkable changes in nucleolar volume (FISCHER; WEBER; SCHÖLM).

Other indications of diurnal variations in the nucleus are circadian differences in ^{14}C-leucine incorporation into protein of liver nuclei in rats (SESTAN).

KLUG reports a diurnal periodicity of the nuclear volume in the corpora allata of *Carabus nemoralis*. Evidently it is related to the circadian rhythm of this animal's activity (see Figs. 40 and 41).

Mitotic cycles. These facts are also interesting in another respect: mitosis may be one of the processes controlled by the circadian rhythm. The small temperature dependence shown in Fig. 102 demonstrates how closely mitosis can be tied to the endodiurnal system. A diurnal rhythm of cell division is known to occur in many other plants and animals, at least under normal LD cycles. Decades ago the continuation of the rhythm in cell division under constant conditions was described. Subsequent observations in animals also showed continuation of the mitotic rhythm under constant conditions (see *HALBERG et al.; UTKIN

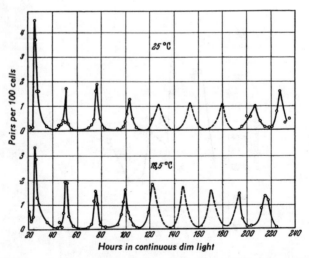

Fig. 102. *Gonyaulax polyedra*. Diurnal rhythm of cell division at 18.5 and at 25°C, as measured by the percentage of paired cells present in a cell suspension. Constant dim light. After HASTINGS and SWEENEY

and KOSICHENKO; VASAMA). The individual organs and tissues in mice differ in their ability to continue these cycles under constant conditions (HALBERG and BARNUM; see also *HALBERG).

> In some plants the restriction of mitosis to the night time is so strict that older botanists, looking for different division stages in certain algae, examined them either during the night or stopped the physiological clock temporarily by cooling it down to temperatures near the freezing point.

But mitosis itself is certainly not an essential part of the clock mechanism. The mitotic cycle *can* be under control of the clock: only certain phases (gates) of the circadian cycle may allow mitosis (see p. 14). EDMUNDS and FUNCH have referred to this gating function of the circadian clock when studying the cell division cycles of *Euglena*.

Studies concerning the rhythm of mating type reversal in *Paramecium* demonstrate very clearly that the circadian cycle may be independent of the division cycle. The circadian cycle is inherited through repeated cell replication, even when the division cycle is considerably shorter than the expressed circadian period (BARNETT).

For these reasons, biochemical studies with cells or tissues showing mitotic activity do not allow far-reaching conclusions about the basis of the circadian clock. The biochemical characteristics within the mitotic cycle may be wrongly interpreted as characteristics of the circadian cycle.

Continuation of the clock in enucleated cells. The observations on the cyclic behavior of nuclei and nucleoli, as well as the inhibiting effects of actinomycin on the circadian rhythmicity (KARAKASHIAN and HASTINGS 1963), may lead to the conclusion that the "master-clock" is located in the nucleus. This cannot, however, be the whole truth. This becomes clear especially from experiments with the alga *Acetabularia*, in which the diurnal rhythm in photosynthetic capacity and in the shape of the chloroplasts continues for several days or even for more than 1 month under constant conditions in cells from which the nucleus has been removed (SWEENEY and HAXO; SCHWEIGER et al.; VANDEN DRIESSCHE).

However, even these experiments do not allow the conclusion that the role of the nucleus in the circadian cycle is only a passive one. Though the rhythmicity continues in enucleated cells of *Acetabularia*, the nucleus is capable of determining the phase of the rhythm (SCHWEIGER et al.). The inhibition of the rhythmicity in *Gonyaulax* by actinomycin (KARAKASHIAN and HASTINGS 1963) is not a decisive argument for the active participation of the nucleus, since DNA-dependent RNA synthesis occurs also in other cell constituents.

g. Role of nucleic acids

Indications for the role of nucleic acids. A participation of nucleic acids is indicated by the phase-shifting effect of ultraviolet mentioned earlier. In *Gonyaulax*, HASTINGS tried to find, without much success, fluctuations in the incorporation of inorganic phosphate into nucleic acids (see *HASTINGS; *HASTINGS and KEYNAN). Studies by *HALBERG *et al.* on the metabolism of DNA and RNA in mammals are quite remarkable. EHRET found with *Paramecium* only slight indications of diurnal fluctuations in nucleic acid metabolism, or in the activity of participating enzymes. Further examples for circadian rhythms in DNA or RNA are discussed by SCHEVING and PAULY; and MERRIT and SULKOWSKI.

Negative findings do not prove with certainty that nucleic acids do not participate in the clock itself, because there is still a possibility that only a small fraction of the total nucleic acid metabolism is necessary for the control of the periodicity, as has been pointed out by *HASTINGS (1960).

Many reports concerning nucleic acid rhythmicity refer to organisms kept in LD cycles and to tissues with dividing cells (IZQUIERDO and GIBBS). Thus, these changes may just be characteristics of the cell cycle, not of the "true" circadian cycle (see Chapter 2(b)).

In *Acetabularia* there seems to be a real circadian RNA cycle. VANDEN DRIESSCHE and BONOTTO found an endogenous rhythm in the incorporation of labeled uridine into the RNA fractions. Enucleated algae show the same rhythm. According to these authors' opinion, the rhythm is associated with the chloroplasts.

Possible role of chloroplasts. In order to understand the presence of the clock in enucleated cells in spite of the evident role played by DNA-dependent RNA synthesis, we may focus our attention on the possible participation of the chloroplasts (SCHWEIGER and SCHWEIGER; SCHWEIGER and BERGER). DNA has been found associated with chloroplasts, and DNA-dependent RNA synthesis has been described in these organelles (*PARTHIER and WOLLGIEHN; SHAH and LYMAN; COOK).

Evidently, the several circadian functions associated with the chloroplasts can be maintained without nuclear control. In addition to photosynthetic capacity, we may mention that the ability to form chlorophyll also fluctuates diurnally (CLAUSS and RAU; CLAUSS and SCHWEMMLE; MITRAKOS). Of course, light is necessary for the formation of chlorophyll, but if the plant is exposed to the same light intensity at different times, the amount of chlorophyll synthesized varies

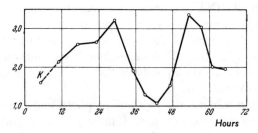

Fig. 103. *Hyoscyamus niger*. Relative amounts of chlorophyll produced in 72-hour cycles. Each cycle consisted of 10 hours' light followed by 62 hours' darkness. During the long dark period an additional light period of 2 hours was given. The time of this additional light period was different in the several experimental sets. K: controls without this light break. Abscissa: time of the 2-hour light break within the 72-hour cycle. The curve shows that the light break influences chlorophyll formation in different degrees and even in a different direction at the several times. Phases with equal physiological character about 24 hours apart. After CLAUSS u. RAU

according to an endogenous diurnal pattern. Sometimes light may favor chlorophyll formation, but at other times it evidently favors breakdown or at least cannot compensate for breakdown (Fig. 103).

The shell structure of starch grains is another expression of endodiurnal fluctuations within the plastids. The starch grains of the potato tuber still show this structure even when the tuber is grown under constant conditions (HESS). Electron microscope studies have confirmed that shell formation in potato and tobacco starch grains (BUTTROSE, 1963) must be controlled by an endogenous rhythm, whereas in wheat it must be controlled by external factors (BUTTROSE, 1962). Since the layers are due to different water content, their formation is also the expression of a qualitative change in enzyme activity.

Finally, diurnal fluctuations in the shape of the plastids, occurring also under constant conditions (see BUSCH), may lead to the assumption that periodical changes in the structure of plastids are the basis for the periodical change in enzyme activity. The parallelism between endogenous diurnal changes in the shape of plastids and in the photosynthetic capacity has been studied with higher plants (BÜNNING 1942) and more recently with the giant cells of *Acetabularia* (VANDEN DRIESSCHE).

Biochemical details on the rhythm in photosynthetic capacity in *Gonyaulax* have been published by SWEENEY (1965). The Hill reaction did not show the rhythmicity, but there were rhythms in the activity of the enzyme that catalyzes the formation of phosphoglyceric acid from ribulose diphosphate and carbon dioxide. (See also HOFFMAN and MILLER.)

Inhibition of RNA synthesis. KARAKASHIAN and HASTINGS succeeded in suppressing the rhythm of photosynthetic capacity in *Gonyaulax* with actinomycin D. Actinomycin is a substance that seems to affect cells by inhibiting DNA-dependent RNA synthesis. This effect of actinomycin favors the hypothesis that the clock function is dependent upon RNA synthesis.

Actinomycin inhibits the rhythms of photosynthetic capacity and of bioluminescence in *Gonyaulax*. Actinomycin also inhibits rhythmic sap exudation from the roots of vascular plants (MACDOWALL). These results indeed indicate a specific action on the clock, especially since the processes normally under control of the clock are not necessarily fully suppressed when their rhythmicity is prevented by actinomycin D (*HASTINGS and KEYNAN).

Other experimental facts are not consistent with the hypothesis that assumes a decisive role for DNA-dependent RNA synthesis. In enucleated cells of *Acetabularia*, actinomycin has no effect on the phase or the period of the photosynthetic rhythm (SWEENEY et al. 1967).

General remarks. How to reconcile these apparently contradictory facts? VANDEN DRIESSCHE tried to explain the inhibiting effect of actinomycin by assuming the participation of a long-lived messenger RNA. In addition to this long-lived RNA, some circadian transcription might be involved.

EHRET and TRUCCO suggested a corresponding model. They proposed a "chronon concept." "The chronon is a very long polycistronic complex of DNA whose transcription rate is limited by some functions of eukaryotic organisation that are relatively temperature independent. Each eucell contains hundreds of chronons on each of its nuclear chromosomes, and many sets of extranuclear chronons . . . . RNA transcription proceeds unidirectionally from the initiator cistron . . . to the terminator cistron."

It is not easy to explain certain experimental facts on the basis of hypotheses assuming circadian RNA syntheses. Not only actinomycin D, but also puromycin and chloramphenicol, though strongly inhibiting protein synthesis, showed no effect on the photosynthetic rhythm of *Acetabularia* (SWEENEY et al.). In the same alga, rifampicin, a specific inhibitor of mitochondrial and chloroplastic RNA polymerase, failed to influence the circadian rhythmicity (VANDEN DRIESSCHE et al. 1970). Therefore, the rhythm is apparently not dependent on a recurrent daily nuclear or chloroplastic transcription.

Finally, we should not forget that actinomycin is known to have also other effects on cells besides inhibiting DNA-dependent RNA synthesis (Example, TANAKA et al.).

h. The microscopic and ultrastructural approach

Diurnal structural changes can be detected both by light and electron microscopy. We mentioned circadian changes in the volume of the nuclei and in the shape of chloroplasts. Ultrastructural studies reveal changes in the endoplasmic reticulum and in the mitochondria. Many of these changes are known to be connected with enzyme rhythms (*MAYERSBACH; CHEDID and NAIR). However, most of these studies refer to conditions with LD cycles. Even in case these structural rhythms should continue in LL or DD, they may be suspected to be consequences, not components of the unknown clockwork. The reason for this statement is the fact that in other species or with other experimental conditions these overt microscopic or submicroscopic changes may be missing, though the clock continues to run.

The unicellular alga *Gonyaulax* shows circadian rhythms in bioluminescence, in photosynthetic capacity, and in several other physiological processes. Cells of *Gonyaulax* were fixed at different times of day in light-dark cycles and in constant light. "Examination of thin sections under the electron microscope . . . failed to show any differences in structure which could be correlated with the time in the cycle when the cells were fixed, although nucleus, nuclear membrane, Golgi bodies, mitochondria, plastids and cell membranes were well preserved" (*SWEENEY 1969).

i. Cooperation of several cell constituents

Different autonomous oscillators within one cell? In order to overcome the difficulties in combining the several experimental facts, one might assume a cooperation of several circadian clocks within the cell. The fact that not only the nucleus but also plastids and mitochondria (*TUPPY and WINTERSBERGER) contain DNA might be a help for such an assumption.

However, even such a hypothesis faces difficulties. In individual cells showing two, three, or more different circadian processes, it was never possible to change by any experimental treatment the phase angle difference between these several processes (MCMURRY and HASTINGS).

Role of mitochondria? BRINKMANN (1971) draws attention to the possible role of mitochondria. One of his arguments is as follows. Bacteria are lacking in mitochondria and do not display circadian rhythms. Fungi possess different types of mitochondria (with different biochemical functions). As a consequence, they may show circadian rhythms, which are only to a certain degree temperature-compensated, they may also reveal rhythms with longer periods. Other plant and animal cells exhibit a similar type of mitochondria, and they show the characteristic temperature-compensated circadian rhythmicity.

The suggested role of mitochondria is consistent with the effect of valinomycin (Fig. 104), since this antibiotic induces especially an accumulation of K^+ by mitochondria (PRESSMAN).

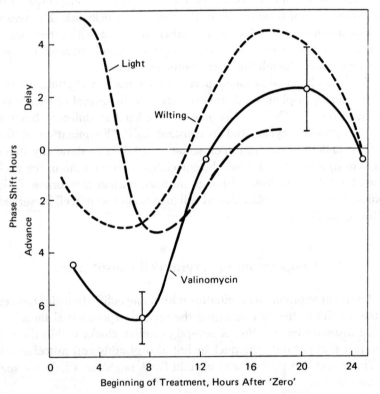

Fig. 104. *Phaseolus coccineus*. Phase shifts of the free-running circadian leaf movements in LL, due to transient wilting (6 hours) and to transient treatment with valinomycin (5 hours, offered via the transpiration stream, 10^{-2} mg/liter water). The treatment started at various phases of the circadian cycle. "Zero" indicates the last subjective night peak of the circadian cycle before treatment. The data refer to delays or advances in the 3rd (wilting) or 4th cycle (valinomycin) after treatment. Data are compared with phase-response curves for white light pulses. From BÜNNING and MOSER 1972

j. The possible role of membranes

General remarks. The experimental facts summarized in the preceding pages do not allow a satisfying theory on the biochemical mechanism of the clock. Apparently, at least most of the known circadian biochemical oscillations are not wheels of the clock, but result from still unknown oscillations. A biophysical approach may be more successful than the biochemical approach. A more detailed knowledge concerning membrane biology might be helpful.

Periodic phenomena in artificial membranes are known. TEORELL found that rhythmic changes can occur in the membrane potentials, membrane resistance, and water flow through membranes caused by electric currents running through salt solutions of different concentrations, provided these solutions were separated from each other by a charged membrane.

> For additional research concerning the periodic behavior of charged membranes see ARANOW. A general review on periodic phenomena in membranes was published by °KATCHALSKY and SPANGLER.

A hypothesis for the clock mechanism based on membrane processes would mean that the great variety of biological rhythms ranging from those with periods of a few milliseconds up to the circadian ones have something in common. All of them are delayed by heavy water (ENRIGHT 1971a, b). And since the "high-frequency rhythms originate in pacemakers dependent on diffusion processes, the experimental results suggest the possibility that long-period biological clocks are also based on diffusion-dependent pacemakers" (ENRIGHT 1971a).

A biophysical approach has also the advantage of bypassing the phenomenon of the limited influence of temperature on the length of the periods.

Role of water. The effects of alcohol and D_2O (see Chapter 8(c)) on the clock are strong arguments for assuming the involvement of membrane processes in the clock mechanism. In addition, we may mention the role of normal water. Withdrawing water can influence phase and period in a similar way as alcohol or D_2O affect the clock. In *Phaseolus,* a continuous wilting may increase the period by 1–3 hours; a 6-hour period of wilting, offered during various phases of the cycle, results in advances or delays (Fig. 104). The phase-response curves resemble the phase response curves for light pulses.

Effect of substances known to affect permeability. In addition to the effects of alcohol and D_2O, we may mention the effects of lithium ions and of valinomycin. Lithium ions are known to affect membrane permeability (ROSE and LOWENSTEIN). They lengthen the period of the *Kalanchoe* clock (ENGELMANN). The antibiotic valinomycin especially causes an increase in the turnover of K^+. In *Phaseolus,* a brief treatment with valinomycin induces phase shifts, which are similar to those induced by temporary wilting or by light pulses (Fig. 104). In isolated eyes of *Aplysia california* high K^+ pulses result in phase shifts (advances or delays) which also are similar to light induced phase shifts (ESKIN).

Circadian changes in permeability. There is now increasing evidence for circadian changes in permeability. Earlier experiments (BÜNNING 1934) indicated that changes in water permeability are a decisive factor in the circadian leaf movements of *Phaseolus*. More recently, SATTER and GALSTON published evidence for the role of circadian potassium fluxes in the circadian leaf movements of *Albizzia*. WAGNER and CUMMING suggest that there is a rhythm in the stability of the cellular membranes. This suggestion is based on experiments concerning betacyanin leakage from seedlings of *Chenopodium rubrum*.

KITTLICK found diurnal changes of membrane permeability in liver cell nuclei of the rat. But in this case it is not known whether the changes continue in LL or DD.

References to Chapter 8

(*Attempts toward a Biochemical and Biophysical Analysis*)

a. Reviews

*GOODWIN, B. C. 1963. *Temporal Organization in Cells*. London–New York: Academic Press.
*HALBERG, F. 1960. Temporal coordination of physiologic function. *Cold Spring Harbor Symp. quant. Biol.* 25:289–308.–
*HALBERG, F., E. HALBERG, C. P. BARNUM, and J. J. BITTNER. 1959. In *Photoperiodism and Related Phenomena in Plants and Animals*. R. B. WITHROW, ed. Washington: Am. Ass. Adv. Sci. Pp. 803–878.–
*HASTINGS, J. W. 1960. Biochemical aspects of rhythms: phase shifting by chemicals. *Cold Spring Harbor Symp. quant. Biol.* 25:131–140.–
*HASTINGS, J. W. 1970. Cellular-biochemical clock hypothesis. In *The Biological Clock: Two Views*. F. A. BROWN, J. W. HASTINGS, and J. D. PALMER, eds. New York–London: Academic Press.–

*HASTINGS, J. W., and V. C. BODE. 1962. Biochemistry of rhythmic systems. Ann. N. Y. Acad. Sci. 117:876–889.–
*HASTINGS, J. W., and A. KEYNAN. 1965. Molecular aspects of circadian systems. In Circadian Clocks. J. ASCHOFF, ed. Amsterdam: North Holland Publishing Company. Pp. 167–182.
*KATCHALSKY, A., and R. SPANGLER. 1968. Dynamic membrane processes. Quaterl. Rev. Biophysics 1:127–175.
*MAYERSBACH, H. v., ed. 1967. The Cellular Aspects of Biorhythms. Heidelberg: Springer-Verlag.
*PARTHIER, B., u. R. WOLLGIEHN. 1966. Nucleinsäuren und Proteinsynthese in Plastiden. In Probleme der biologischen Reduplikation. P. SITTE, ed. Berlin: Springer-Verlag. Pp. 244–270.
*SWEENEY, B. M. 1969. Rhythmic Phenomena in Plants. London–New York: Academic Press.
*TUPPY, H., u. E. WINTERSBERGER. 1966. Mitochondrien als Träger genetischer Information. In Probleme der biologischen Reduplikation. P. SITTE, ed. Berlin: Springer-Verlag. Pp. 325–335.
*VAN PILSUM, J. F., and F. HALBERG. 1964. Transamidase activity in mouse kidney–an aspect of circadian enzyme activity. Ann. N. Y. Acad. Sci. 117:337–353.

b. Other references

ARANOW, R. H. 1963. Proc. Nat. Acad. Sci. USA 50:1066–1070.
BALL, N. G., and I. J. DYKE. 1956. J. exp. Bot. 7:25–41;–
———. 1957. J. exp. Bot. 8:339–347.–
BARNETT, A. 1966. J. Cell. Physiol. 67:239–270.–
BERNHARDT, W., K. PANTEN, u. H. HOLZER. 1964. Angew. Chemie 76:990–991.—
BETZ, A., and B. CHANCE. 1965. Arch. Bioch. Biophys. 109:585–594.–
BIEBL, R., u. K. HOFER. 1966. Radiation Biology 6:225–250.–
BRINKMANN, K. 1966. Planta (Berl.) 70:344–389.–
BRINKMANN, K. 1971. In Biochronometry. M. MENAKER, ed. Washington, D.C.: Nat. Acad. Sci. Pp. 567–593.–
BRUCE, V. G., and C. S. PITTENDRIGH. 1960. J. cell. comp. Physiol. 56:25–31.–
BÜHNEMANN, F. 1955. Biol. Zentralbl. 74:691–705.–
BÜNNING, E. 1934. Jahrb. wiss. Bot. 79:191–230;–
———. 1942. Z. Bot. 37:433–486;–
———. 1956. Z. Bot. 44:515–529;–
———. Planta (Berl.) 48:453–458.–
BÜNNING, E., u. J. BALTES. 1962. Naturwiss. 49:19.–
BÜNNING, E., S. KURRAS, u. V. VIELHABEN. 1965. Planta (Berl.) 64:291–300.–
BÜNNING, E., u. I. MOSER. 1968. Naturwiss. 55:450–451;–
———. 1972. Proc. Nat. Acad. Sci. USA 69:2732–2733.–

BÜNNING, E., u. G. SCHÖNE-SCHNEIDERHÖHN. 1957. *Planta* (*Berl.*) **48**:459–467.–
BÜSCH, G. 1953. *Biol. Zentralbl.* **72**:598–629.–
BUTTROSE, M. S. 1962. *J. Cell. Biol.* **14**:159–167;–
———. 1963. *Naturwiss.* **50**:450–451.
CHANCE, B., R. R. W. ESTABROOK, and A. GHOSH. 1964. *Proc. Nat. Acad. Sci. USA* **51**:1244–1251.–
CHANCE, B., and M. NISHIMURA. 1960. *Proc. Nat. Acad. Sci. USA* **46**:19–24.–
CHANCE, B., B. SCHÖNER, and S. ELSAESSER. 1965. *J. Biol. Chem.* **240**:3170–3181.–
CHEDID, A., and V. NAIR. 1972. *Science* **175**:176–179.–
CLAUSS, H., u. W. RAU. 1956. *Z. Bot.* **44**:437–454.–
CLAUSS, H., u. B. SCHWEMMLE. 1959. *Z. Bot.* **47**:226–250.–
COOK, J. 1966. *J. Cell Biol.* **29**:369–373.
EDMUNDS, L. N. 1965. *J. Cell. comp. Physiol.* **66**:159–181;–
———. 1966. *J. Cell. comp. Physiol.* **67**:35–43.–
EDMUNDS, L. N., and R. R. FUNCH. 1969. *Planta* (*Berl.*) **87**:134–163;–
———. 1969. *Science* **165**:500–503.–
EHRENBERG, M. 1950. *Planta* (*Berl.*) **38**:244–279.–
EHRET, C. F. 1960. *Cold Spring Harbor Symp. quant. Biol.* **25**:149–157.–
EHRET, C. F., and E. TRUCCO. 1967. *J. theor. Biol.* **15**:240–262.–
ENGELMANN, W. 1972. *Z. Naturforsch.* **27b**:477.–
ENRIGHT, J. Y. 1971a. *Z. vergl. Physiol.* **72**:1–16;–
———. 1971b. *Z. vergl. Physiol.* **75**:332–346.
ESKIN, A. 1972. *J. comp. Physiol.* **80**:353–376.
FELDMAN, J. F. 1967. *Proc. Nat. Acad. Sci. USA* **57**:1080–1087;–
———. 1968. *Science* **160**:1454–1456.–
FISCHER, H. 1934. *Planta* (*Berl.*) **22**:767–793.
GLICK, J. L., and W. D. COHEN. 1964. *Science* **143**:1184–1185.–
GLICK, D., R. B. FERGUSON, L. J. GREENBERG, and F. HALBERG. 1961. *Amer. J. Physiol.* **200**:811–814.–
GRABENSBERGER, W. 1933. *Z. vergl. Physiol.* **20**:1–54.
HALBERG, F., and C. P. BARNUM. 1961. *Amer. J. Physiol.* **201**:227–230.–
HARDELAND, R. 1969. *Z. vergl. Physiol.* **63**:119–136.–
HASTINGS, J. W., and B. M. SWEENEY. 1957. *Proc. Nat. Acad. Sci. USA* **43**:804–811.–
HESS, C. 1955. *Z. Bot.* **43**:181–204.–
HOFER, K., u. R. BIEBL. 1965. *Protoplasma* **59**:506–521.–
HOFFMAN, F. M., and J. H. MILLER. 1966. *Am. J. Bot.* **53**:543–548.
IZQUIERDO, J. N., and S. J. GIBBS. 1972. *Exp. Cell Res.* **71**:402–408.
JACKLET, J. W., and J. GERONIMO. 1971. *Science* **174**:299–302.
KALMUS, H. 1934. *Z. vergl. Physiol.* **20**:405–419.–
KARAKASHIAN, M. W., and J. W. HASTINGS. 1962. *Proc. Nat. Acad. Sci. USA* **48**:2130–2137;–
———. 1963. *J. gen. Physiol.* **47**:1–12.–

KELLER, S. 1960. *Z. Bot.* **48**:32–57.–
KITTLICK, P.-D. 1970. *Exp. Path. (Jena)* **4**:143–154; 204–206.–
KLITZING, L. v. 1969. *Protoplasma* **68**:341–350.–
KLUG, H. 1958. *Naturwiss.* **45**:141–142.
LE BOUTON, A. V., and S. D. HANDLER. 1971. *Experientia* **27**:1031–1032.–
MACDOWALL, F. D. H. 1964. *Canad. J. Bot.* **42**:115–122.–
MCMURRY, L., and J. W. HASTINGS. 1972. *Science* **175**:1137–1139.–
MASTERS, S., and W. D. DONACHIE. 1966. *Nature* **209**:476–479.–
MERRIT, J. H., and T. S. SULKOWSKI. 1970. *J. Neurochem.* **17**:1327–1328.–
MITRAKOS, K. 1959. *Planta (Berl.)* **52**:583–586.–
MÖDLINGER-ODOREER, M. 1962. *Endokrinologie* **43**:45–60.
NIEBROJ, T. 1958. *Naturwiss.* **45**:67.
PALMER, J. D., and H. B. DOWSE. 1969. *Biol. Bull.* **137**:388.–
PAVLIDIS, T. 1969. *J. theor. Biol.* **22**:418–436;–
———. 1971. *J. theor. Biol.* **33**:319–338.–
PAVLIDIS, T., and W. KAUZMANN. 1969. *Arch. Bioch. a. Biophys.* **132**:338–348.–
PRESSMAN, B. C. 1968. *Federation Proc.* **27**:1283–1288.
QUAY, W. B., and A. RENSONI. 1966. *Growth* **30**:315–324.
RENNER, M. 1957. *Z. vergl. Physiol.* **40**:85–118;–
———. 1958. *Ergebn. Biol.* **20**:127–158.
RICHTER, G., u. A. PIRSON. 1957. *Flora* **144**:562–597.–
ROSE, B., and W. R. LOWENSTEIN. 1971. *J. Membr. Biol.* **5**:20–50.–
SATTER, R. L., and A. W. GALSTON. 1970. *Science* **174**:518–519;–
———. 1971. *Plant Physiol.* **48**:740–746.–
SCHEVING, L. E., and I. E. PAULY. 1967. *Cell Biol.* **32**:677–683.–
SCHÖLM, H. E. 1968. *Protoplasma* **66**:393–401.–
SCHWEIGER, E., and S. BERGER. 1964. *Bioch. Biophys. Acta.* **87**:533–535.–
SCHWEIGER, H. G., and E. SCHWEIGER. 1965. In *Circadian Clocks*. J. ASCHOFF, ed. Amsterdam: North Holland Publishing Company. Pp. 195–197.–
SCHWEIGER, E., H. G. WALLRAFF, u. H. G. SCHWEIGER. 1964. *Z. Naturforsch.* **19b**:499–505.–
SESTAN, N. 1964. *Naturwiss.* **51**:371.–
SHAH, V. C., and H. LYMAN. 1966. *J. Cell Biol.* **29**:174–176.–
STEINHEIL, W. 1970. *Z. Pflanzenphysiol.* **62**:204–215.–
SULZMAN, F. M., and L. N. EDMUNDS. 1972. *Bioch. Biophys. Res. Communications* **47**:1338–1344.
SUTER, R. B., and K. S. RAWSON. 1968. *Science* **160**:1011–1014.–
SWEENEY, B. M. 1965. In *Circadian Clocks*. J. ASCHOFF, ed. Amsterdam: North Holland Publishing Company. Pp. 190–194;–
———. 1964. *Plant Physiol. Suppl.* **39**:14 (abstr.);–
———. 1969. *Canad. J. Bot.* **47**:299–308;–
SWEENEY, B. M., and F. T. HAXO. 1961. *Science* **134**:1361–1663.–
SWEENEY, B. M., C. F. TUFFLI, and R. H. RUBIN. 1967. *J. gen. Physiol.* **50**:647–659.

TANAKA, Y., H. F. DE LUCA, J. OMDAHL, and M. F. HOLICK. 1971. *Proc. Nat. Acad. Sci. USA* **68**:1286–1288.–

TEORELL, T. 1959. *J. gen. Physiol.* **42**:831–845;–

———. 1962. *Biophys. J.* **2**:27–52.–

TSCHUDY, P. D., A. WAXMAN, and A. COLLINS. 1967. *Proc. Nat. Acad. Sci. USA* **58**:1944–1948.

UTKIN, I. A., and L. P. KOSICHENKO. 1961. *Doklady Akad. Nauk. SSSR.* **134**:119–194.

VANDEN DRIESSCHE, TH. 1966. *Exp. Cell Res.* **42**:18–30.–

VANDEN DRIESSCHE, TH., and S. BONOTTO. 1969. *Bioch. Biophys. Acta* **179**:58–66.–

VANDEN DRIESSCHE, TH., S. BONOTTO, and J. BRACHET. 1970. *Bioch. Biophys. Acta* **224**:631–634.–

VASAMA, R. 1961. *Ann. Univers. Turkuensis Ser.* A. II. Nr. 29.–

VENTER, J. 1956. *Z. Bot.* **44**:59–76.

WAGNER, E., and B. G. CUMMING. 1971. *Canad. J. Bot.* **48**:1–18.–

WEBER, F. 1926. *Planta (Berl.)* **1**:441–471.–

WERNER, G. 1957. *Z. vergl. Physiol.* **36**:464–487.–

WILKINS, M. B. 1967. *Planta (Berl.)* **72**:66–77.–

WILKINS, M. B., and D. M. WARREN. 1963. *Planta (Berl.)* **60**:261–273.

9. Adjustment to Diurnal Cycles in the Environment

a. Synchronization with physical rhythms of the environment

The circadian clock enables plants and animals to adjust themselves to diurnal changes of external factors, i.e., to perform certain activities at a time of day most suitable for that particular activity. Since this properly falls in the realm of ecology (*CLOUDSLEY-THOMPSON 1960, 1970; *REMMERT 1969), only a few details will be mentioned to illustrate the possibilities.

*WENT has stressed the ecological implications of the circadian rhythms in plants. The daily cycles of light and darkness as well as the daily cycles of temperature are important. Though we shall return to this point when discussing pathological effects (Chapter 14), here are a few examples.

Tomatoes, grown in continuous light, show maximum increase in dry weight when the temperature has been fluctuating daily between 26°C (for 16 hours) and 10°C (for 8 hours) (KRISTOFFERSEN). The length of the LD cycle in which tomato, pea, pea nut, and soy-bean plants achieve maximum growth is close to 24 hours for cycles consisting of equal periods of light and darkness (TUKEY and KETELLAPPER).

Light-dark cycles. The importance of organisms' being able to adjust themselves to the alternation of day and night is quite obvious. Consider, for example, the controlling influence of the circadian clock on changes of activity and inactivity of sleep and wakefulness. In plants, the high photosynthetic capacity during the daytime and more intense dissimilatory abilities during the night (see Chapter 2(a)) should also be mentioned.

Cycles of high and low temperatures. The temperature optimum of organisms can be controlled by the clock in a diurnal pattern. In plants, it is usually lower at night than during the day. Most plants develop normally only if the night temperature is lower than the daytime temperature (so-called thermoperiodism, see *WENT; *SCHWEMMLE).

TAYLOR gives an instructive example of the advantage accruing to certain animals that are able to adjust to the change of high and low temperatures. If the lizard *Sceloporus magister* is exposed to diurnal cycles of 20° and 35°C, it buries itself in sand when the temperature becomes low and emerged when the temperature rises. After a few days, it buries itself shortly before the temperature decreases. This behavior was maintained for a week after the temperature was kept constantly high.

Cycles of high and low humidity. Some insects emerge from their pupae at a certain time of day. This diurnal fixation is often an adjustment to humidity changes. It can be advantageous to emerge at daybreak (often shortly before sunrise, as *Drosophila*, for example, do) since the freshly emerged insects are not exposed immediately to dehydration by sunlight, which may be damaging. Just after emerging their cuticula may not be waterproof (see *REMMERT 1962). Some other adaptations to humidity changes have also been described (*CLOUDSLEY-THOMPSON).

In all these cases it is advantageous to find the appropriate time by diurnal oscillations, because the organism can "plan." For example, such inhibiting factors as low temperature may restrict the physiological process in question (e.g., emergence from the pupae, photosynthesis), but this does not impair the ability to find the appropriate time on the following days.

b. Synchronization with biological rhythms of the environment

Provisions for sexual mating. In some animals and plants the readiness for sexual mating depends on physiological conditions, which obtain only during a few hours of the day. These conditions may be controlled by the circadian clock. If the readiness is not synchronized in individuals of different sex, sexual union will be impossible. The reactions of *Paramecium* (Chapter 4(a)) might be quoted here (SONNEBORN; EHRET; KARAKASHIAN). In many species of moths females release pheromones which attract males over long distances. This release as well as the reactivity to the pheromones can be controlled by the circadian clock and thus be restricted to a few hours of the day. This rhythm may persist in LL (SOWER *et al.*).

Prey catching. Synchronizing the periodicity of predatory animals with the periodicity of the prey can become an important factor for catching of the prey. For example, the jumping of salmon in the evening is synchronized with the maximum activity of the hunted insects (see *Aschoff).

Insects visiting flowers. The synchronization of the "flower clock" with the clock of insects was mentioned in the introduction. Bees and other insects are able to learn the time at which the different kinds of flowers offer pollen. The processes controlled by the endogenous diurnal rhythm are not only closing and opening of the flowers or secretion of nectar, but also the production of odor. The investigations on flowers by Overland quite clearly show that this can take place in an endodiurnal pattern.

c. Special questions on the time sense of insects

Some historical facts. Forel first made the striking observation that bees visited his marmalade regularly at breakfasttime. Then he experimentally proved the assumption that the bees were guided by a sense of time. It had been noticed quite early that the bees came to certain flowers at a certain time, and that this coincided with the secretion of nectar, which is also bound to certain times of day.

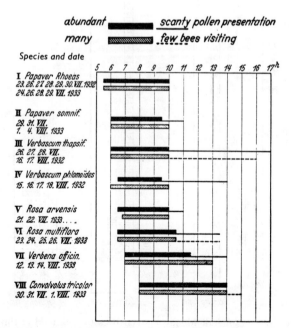

Fig. 105. Times of day when flowers present pollen and when they are visited by bees. After Kleber.

KLEBER has published detailed observations on the diurnal pattern of nectar secretion in many flowers (Fig. 105).

During investigations at v. FRISCH's laboratory on the time sense of bees BELING examined these phenomena more closely than anyone had done before.

Phenomena. The time sense is not based on a learning of intervals. Bees cannot be trained to an interval deviating from the 24-hour periodicity such as, say, an interval of 19, 27, or 48 hours. It is possible, however, to train them to several times of day (WAHL, Fig. 106). They remember the training times even if they were trained to 9 different times of day, and they are able to distinguish between two points in time that are separated by not more than 20 minutes (KOLTERMANN).

It is especially interesting to notice that the bees always visit the right place at a given training time.

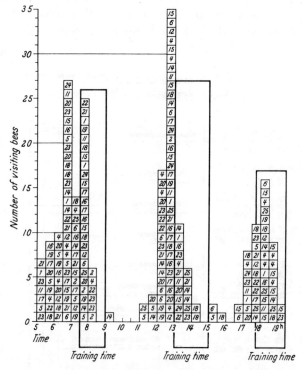

Fig. 106. Training of individually marked bees to seek food at three different times of the day. The bees were trained for 6 days. The visits were recorded on the 7th day when no food was provided. Compare with Fig. 1. After BELING

*RENNER, by performing a transoceanic experiment, has removed all doubt that this phenomenon is in fact based on an endogenous clock. Bees were trained to a certain time of day in the European time zone. After being flown to a different American time zone, their behavior was examined. It corresponded to the original time zone. The training to a certain time of day can be maintained for 6 to 8 days (WAHL). BEIER u. LINDAUER demonstrated the circadian nature of this endogenous clock: in LL, the period is about 23.4 hours.

These phenomena of time sense in bees and other insects lead to the following question: How does the animal, metaphorically speaking, mark a certain time on its clock (as we do with a rider on the dial of a time switch)? The simplest answer would be that the clock shifts its phase position because of the synchronizer—"food presentation." This answer has to be rejected since other diurnal functions in bees continue with the old phase position. Moreover, bees can be trained to several times of day. It would be premature to form a hypothesis on the basis of these facts. But a certain analogy to phenomena in plants is noticeable. Plants, too, are able to "retain" certain times of day. If we expose them to a short light stimulus (or certain other stimuli, e.g., temperature shock) this may lead to a notch on the curve of the diurnal leaf movements. On the following days this will again appear at the same time of day. The placing of several "marks" within a 24-hour cycle is also possible. If this is a valid analogy, it means that the bee does not set several tabs on the dial of *one* time switch—again metaphorically speaking—but uses several parallel clocks, localized in the individual elements of the nervous system. This interpretation agrees with the fact that in organs of mammals the different parts of the same individual can show individual cycles, which run parallel to each other, yet with shifted phases (see Chapter 4(a) and Chapter 14).

MEDUGORAC u. LINDAUER (also BEIER *et al.*) brought direct evidence for this interpretation. The time memory of bees was shown to disintegrate during a deep and long lasting CO_2 narcosis. After the narcosis the bees appeared at the feeding place not only during the time of day to which they had been trained before, but (the same individuals) once more a few hours later.

d. Action directed to a preset time by the "head clock"

The human head clock has hardly been investigated experimentally, although it has been known for centuries. The head clock enables us to wake up at a preset time from normal sleep and also from hypnosis with remarkable punctuality (Fig. 2). This is coupled with other manifestations. Even after 3 or more days some people are still able to estimate time with a deviation of less than 1%, although external cues have been experimentally excluded (MacLeod and Roff). This accuracy is no longer surprising to us since we know that the circadian clock in birds and rodents is able to work with a deviation of only a few minutes per day (Chapter 2(d)). Even some poikilothermic animals and plants show a degree of variation of only 15–30 minutes. Averaged over a period of several months, some plants can distinguish differences in day length of merely one minute (an example of this is the relation between day length and rate of development in the rice plant, see Chapter 13(b)).

One of the most striking accomplishments of the head clock is the carrying out of an action ordered by the hypnotist to be performed at a certain time (e.g., "after 3 hours and 50 minutes").

The head clock does not simply give an alarm signal at the predetermined time. Rather, at the preset time, one becomes conscious of the purpose for the alarm signal, "one does not have to stop and think, but one simply knows the reason for waking up, what the intention was, and what it was all about." This is *Clauser's conclusion from his numerous experimental data.

> Clauser describes in his treatise the relation between endogenous diurnal activity periods in animals and the efficiency of the human head clock. Clauser mentions the case of an orangutan being transported by ship from Java to Europe. The orangutan always slept exactly 12 hours but also gradually shifted his clock (as we noted for the endogenous diurnal rhythm). The animal did not adjust with a rate corresponding to the speed of the ship, however, but slept from 2 p.m. until 2 a.m. when the ship passed the meridian of the Cape of Good Hope (instead of sleeping from 6 p.m. to 6 a.m., as it would have in Java).

e. The synchronizing factors

Known synchronizers. Different factors can serve as synchronizers ("Zeitgeber") for the adjustment to the diurnal processes of the environment. The adjustment to the rhythm of another organism is often possible when both rhythms are controlled by the same external factor (e.g., LD cycle); but an adjustment to another rhythm is also possible when the other rhythm itself serves as a synchronizer. Besides the LD cycle, we also became acquainted with the synchronizing effect of alternating high and low temperature. The rhythmic change of tides can also function as a synchronizer, as can the fluctuations of humidity, although in rare cases (CLOUDSLEY-THOMPSON). Social synchronizers in the fullest sense of the word are especially important for animals and humans. In humans, they are even more effective than an artificial LD cycle (WEVER 1971).

*ASCHOFF has summarized the available experimental data on synchronizers. He also took into consideration the competition of several synchronizers. In higher animals and in humans the number of possible synchronizers is so large because of the many important correlations to the environment. Thus, it is often difficult to demonstrate the endogenous rhythm as clearly in higher animals and humans as it is in plants and in lower animals.

There is a variety of possible and often-overlooked synchronizers (e.g., minute sound simuli, and also social or psychic factors). This may explain why the "inversion experiment" in higher animals and humans (light during the night and darkness during the day) sometimes does not result in a continuing phase-shifted periodicity after a transfer to DD or LL. "Spontaneous" relapse to the normal phase position may seem to occur (as mentioned in Chapter 6(b)), other causes might also be responsible for such a relapse).

For synchronization by sound, see GWINNER (birds), MENAKER and ESKIN (birds), and MEYER (hamsters). For synchronization by social cues in human beings, see ASCHOFF et al.; and POPPEL. Experiments by WEVER (1967) show some influence of electric fields (see also DOWSE and PALMER; and ROBERTS 1969).

Unknown synchronizers? Frequently, some less conspicuous external factors are proposed as possible synchronizers—for example, fluctuations in atmospheric pressure or fluctuations in the intensity of cosmic radiation (see BROWN et al.).

The possibility of synchronizers such as these can, of course, not be excluded. But before accepting them one should take into account certain sources of error, which have not so far been considered. For example, red light of very low intensity and duration can shift the phases in plants. Light signals of one second or even small fractions of a second (if the intensity is higher) can also be very effective synchronizers in animals. In plants and animals the known synchronizers can set the phases within a few days.

Behavior at the pole and in space. It has often been claimed that diurnal fluctuations of subtle geophysical factors might be unnoticed signals for the organisms, and thus responsible for all circadian rhythms. HAMNER et al. and coworkers have supplied very clear evidence against such a hypothesis: higher and lower plants, as well as mammals and insects continue with their diurnal behavior even at the South Pole, where the rotation of the earth cannot cause any fluctuation in the environment. Humans continue with their circadian rhythms even during space flights (see STRUGHOLD; ALTUKHOV et al.; HALBERG et al. 1970).

Indeed, the occurrence of rhythms of *exactly 24 hours* under conditions of exclusion of all known cues would be a strong argument for the hypothesis of control by quite unknown external factors (BROWN et al.; TERRACINI and BROWN; BROWN 1969). But in most cases observations of accurate 24-hour periods either proved to be due to the influence of neglected factors, or to be the result of the methods applied for frequency analysis (ENRIGHT).

References to Chapter 9

(*Use of the Clock for Adjustment to Diurnal Cycles in the Environment*)

a. Reviews

*ASCHOFF, J. 1954. Zeitgeber der tierischen Tagesperiodik. *Naturwiss.* 41: 49–56.
*BROWN, F. A. 1960. Response to pervasive geophysical factors and the biological clock problem. In *CHOVNICK, pp. 57–71;–
———. 1965. A unified theory for biological rhythm. In *Circadian Clocks.* J. ASCHOFF, ed. Amsterdam: North Holland Publishing Company. Pp. 231–261.
*CHOVNICK, A., ed. 1960. *Biological Clocks.* Cold Spring Harbor Symp. quant. Biol. 25.
*CLAUSER, G. 1954. *Die Kopfuhr.* Stuttgart: F. Enke.–
*CLOUDSLEY-THOMPSON, J. L. 1960. Adaptive functions of circadian rhythms. In *CHOVNICK (ed.), pp. 345–355;–
———. 1961. *Rhythmic Activity in Animal Physiology and Behavior.* New York–London: Academic Press;–
———. 1970. Recent work on the adaptive functions of circadian and seasonal rhythms in animals. *J. interdiscipl. Cycle Res.* 1:5–19.
*REMMERT, H. 1962. *Der Schlüpfrhythmus der Insekten.* Wiesbaden: Franz Steiner;–
———. 1969. Tageszeitliche Verzahnung der Aktivität verschiedener Organismen. *Oecologia* 3:214–266.–
*RENNER, M. 1958. Der Zeitsinn der Arthropoden. *Ergebn. Biol.* 20:127–158;–
———. 1960. The contribution of the honey bee to the study of time-sense and astronomical orientation. In *CHOVNICK (ed.), pp. 361–367.–
*SCHWEMMLE, B. 1960. Thermoperiodic effects and circadian rhythms in flowering plants. In *CHOVNICK (ed.), pp. 239–243.

*WENT, F. W. 1959. The periodic aspect of photoperiodism and thermoperiodicity. In *Photoperiodism and Related Phenomena in Plants and Animals.* R. B. WITHROW, ed. Washington: Am. Ass. Adv. Sci. Pp. 551-564;—
———. 1960. Photo- and thermoperiodic effects in plant growth. In *CHOVNICK (ed.), pp. 221-230;—
———. 1962. Ecological implications of the autonomous 24-hour rhythm in plants. *Ann. New York Acad. Sci.* **98**:866-875.
ALTUKHOV, G., V. VASIL'EV, V. E. BELAI, and A. D. EGOROV. 1965. Ak.

b. Other references

Nauk. S. S. S. R. ser. Biol. Pp. 182-187.—
ASCHOFF, J. M. FATRANSKA, and H. GIEDKE. 1971. *Science* **171**:213-215.
BEIER, W., u. M. LINDAUER. 1970. *Apidologie* **1**:5-28.—
BEIER, W., I. MEDUGORAC, et M. LINDAUER. 1968. *Ann. Epiphyties* **19**:133-144.—
BELING, I. 1929. *Z. vergl. Physiol.* **9**:259-338.—
BROWN, F. A. 1959. *Science* **130**:1535-1544;—
———. 1969. *Canad. J. Bot.* **47**:287-298.—
BROWN, F. A., M. F. BENNETT, and C. L. RALPH. 1955. *Proc. Soc. exp. Biol. (N.Y.)* **89**:332-337.—
BROWN, F. A., Y. H. PARK, and J. R. ZENO. 1966. *Nature* **211**:830-833.—
BROWN, F. A., H. M. WEBB, and M. F. BENNETT. 1958. *Amer. J. Physiol.* **195**:237-243.—
BROWN, F. A., H. M. WEBB, and E. J. MACEY. 1957. *Biol. Bull.* **113**:112-119.—
CLOUDSLEY-THOMPSON, J. L. 1956. *J. exp. Biol.* **33**:576-582.
DOWSE, H. B., and J. D. PALMER. 1969. *Nature* **222**:564.
EHRET, CH. 1959. *Fed. Proc.* **18**:1232-1240;—
———. 1960. *Cold Spring Harbor Symp. quant. Biol.* **25**:149-158.—
ENRIGHT, T. J. 1965a. *J. theor. Biol.* **8**:426-468;—
———. 1965b. In *Circadian Clocks.* J. ASCHOFF, ed. Amsterdam: North Holland Publishing Company. Pp. 31-42.
FOREL, A. 1910. *Das Sinnesleben der Insekten.* München: Reinhrardt.
GWINNER, E. 1966. *Experientia* **22**:765.
HALBERG, F., C. VALLBONA, L. F. DIETLEIN, J. A. RUMMEL, C. E. BERRY, G. C. PITTS, and S. A. NUNNELY. 1970. *Space Life Sci.* **2**:18-22.
HAMNER, K. C., J. C. FINN, G. S. SIROHI, T. HOSHIZAKI, and B. H. CARPENTER. 1962. *Nature (Lond.)* **195**:476-480.
KARAKASHIAN, M. W. 1968. *J. Cell. Physiol.* **71**:197-210.—
KLEBER, E. 1935. *Z. vergl. Physiol.* **22**:221-262.—
KOLTERMANN, R. 1971. *Z. vergl. Physiol.* **75**:49-68.—
KRISTOFFERSEN, T. 1963. *Physiol. Plantarum Suppl.* **1**:1-98.
MACLEOD, R. B., and M. F. ROFF. 1936. *Acta Psychol. Hague* **1**:389-423.—

MEDUGORAC, I., u. M. LINDAUER. 1967. Z. vergl. Physiol. 55:450–474.–
MENAKER, M., and A. ESKIN. 1966. Science 154:1579–1581.–
MEYER, A. 1968. Naturwiss. 55:234–235.
OVERLAND, L. 1960. Amer. J. Bot. 47:378–382.
POPPEL, E. 1968. Pflügers Arch. Ges. Physiol. 299:364–370.
ROBERTS, A. M. 1969. Nature 223:639.
SONNEBORN, T. M. 1938. Proc. Amer. Phil. Soc. 79:411–433.–
SOWER, L. L., H. H. SHOREY, and L. K. GASTON. 1970. Ann. Ent. Soc. Am. 63:1090–1992.–
STRUGHOLD, H. 1965. Ann. N. Y. Acad. Sci. 134:413–422.
TAYLOR, J. L. 1962. Atti VII. Conf. Soc. Ritmi Biol. Siena 1960. Panminerva Medica Torino. P. 153.–
TERRACINI, E. D., and F. A. BROWN. 1962. Physiol. Zool. 35:27–37.–
TUKEY, H. B., and H. J. KETELLAPPER. 1963. Am. J. Bot. 50:110–115.
WAHL, O. 1932. Z. vergl. Physiol. 16:529–589.–
WEVER, R. 1967. Z. vergl. Physiol. 56:111–128.–
WEVER, R. 1971. Pflügers Arch. Europ. Journ. Physiol. 321:133–142.

10. Use of the Clock in Direction Finding

a. Basic phenomena

General remarks. Higher and lower animals are able to use a sun-compass "correctly," that is, they take into account the changing azimuth of the sun during the day. This ability has recently been studied in greater detail. The pioneering work by v. FRISCH (1950) with bees and by KRAMER (1953) with birds should especially be mentioned.

> BRUN recognized this kind of orientation in ants. With a light-tight box, he prevented the animals from continuing their way for several hours. Individuals of *Lasius niger* continued on their way after being released without accounting for the changed position of the sun. *Formica rufa*, however, maintained its original direction of movement, thereby showing that it was able to take into account the movement of the sun which had occurred while the animal was inside the box.

Since the examination of the inner clock used in orientation with the sun compass has shown that it has similar characteristics to those observed in other experiments, we have no reason to imagine another principle applying specifically only to this case.

Animals use a sun-*azimuth* compass in almost all cases which have been closely investigated, i.e., only the horizontal projection of the sun's direction is considered. Very few animals react in a way which indicates that the altitude of the sun offers additional usable information. The basic principle of orientation by the sun and the reactions to "artificial suns" are given in Fig. 107. For recent summaries, see *AUTRUM, *MATTHEWS; *GALLER *et al*.

Vertebrates. HOFFMANN (1954) trained starlings to a certain direction of the compass by the position of the feeding place. He then exposed the animals to an LD cycle which was shifted by 6 hours compared with the original day-night cycle. Thus, he could shift the clock within a few days so that the birds now took off to search for food in a direction deviating by 90° from the training direction (Fig. 108). The deviation occurs to the right or the left, depending upon whether the clock has been advanced or retarded by 6 hours. Several

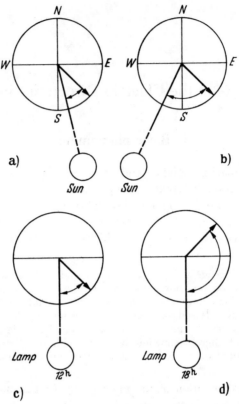

Fig. 107. Physiological clock and sun-compass orientation. An animal is searching for the SE direction. This may be the homing direction or the direction to a feeding place. (a) and (b) the animal chooses different angles to the sun depending on the time of day. Double arrow shows this angle, single arrow the choice of direction. (c) and (d) corresponding behavior without sunlight. The animal changes the angle with the static artificial light source during the day.

days were necessary for adjustment. All of this had been observed earlier in other physiological processes controlled by the clock.

The right direction was still chosen after 28 days of LL and constant temperature, although the rhythm of flying activity was very disturbed by then (°HOFFMANN).

The orientation for the return to home base must be much more difficult. It would require a time sense with an estimated accuracy of 2 minutes per day (KRAMER). We remember that the circadian rhythm in some animals works with deviations of only a few minutes per day.

Experiments by SCHMIDT-KOENIG (1958–1970) with an experi-

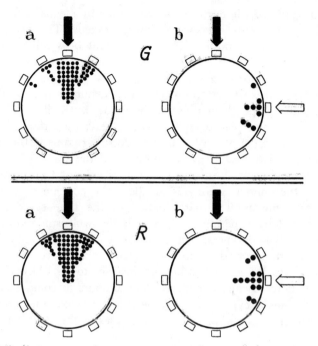

Fig. 108. Compensation for sun movement in direction finding of two starlings. The black arrows indicate the direction of training by offering food before the test. Rectangles represent the (now-empty) food containers. Each dot represents one choice. (a) shows for starling G and R the direction finding after the training period without resetting the clock. (b) shows the choices after the internal clock of the birds had been advanced by 6 hours. Advancing the clock by offering LD cycles with 6-hour phase shift for 12–18 days. White arrows indicate expected choice of direction. After HOFFMANN 1953

mentally phase-shifted clock (as in the above-mentioned experiments of HOFFMANN) indicate that the same kind of orientation does in fact participate in the return of carrier pigeons as in the directed take-off.

For additional work on pigeons, see SCHMIDT-KOENIG (1963–1970); °MATTHEWS; KEETON; SCHMIDT-KOENIG and McDONALD; PAPI and PARDI 1968; PAPI et al. 1971; MICHENER and WALCOTT; and MISELIS and WALCOTT. What kind of other factors are involved in the orientation of pigeons is still a subject of much controversy and not germane to this book (see °GALLER et al.).

Experimental data on the ability of many other vertebrates to make use of time compensated sun-compass orientation are now available: for mammals (LÜTERS u. BIRUKOW), for fishes (BRAEMER; HASLER; SCHWASSMANN and BRAEMER; °HASLER and SCHWASSMANN; °BRAEMER u. SCHWASSMANN), for lizards (°BIRUKOW et al. 1963; FISCHER; FISCHER u. BIRUKOW), for a turtle (FISCHER 1964), and for a frog (FERGUSON et al.).

Arthropods. A similar type of orientation by means of the internal clock has been found in several lower animals. The investigations from v. FRISCH's laboratory demonstrated that bees follow this principle. Special reference has to be made to the experiments of PAPI with the wolf spider (*Arctosa perita*). This animal lives on the banks of rivers and lakes. If placed on the water, it hurries back to the bank in a direction perpendicular to the shore line. But if it is taken to the opposite bank and placed on the water there, then the animal runs across the water (or at least it tries to do so). It runs in the direction to which it was adjusted by the position of the sun. That the angle toward the sun must be decisive becomes evident from experiments in which the animals were misled about the position of the sun by means of a mirror. The participation of the time sense has been demonstrated by experiments in which the wolf spiders were temporarily kept in darkness. The readjustment of these animals (e.g., after being transported to another place) takes more than 1 week. Within 3 days the animals adjust themselves to an LD cycle that is experimentally advanced or delayed by 6 hours in comparison to the normal day-night cycle. So the same rules, derived from other facts, also apply in this case for setting the phases of the circadian rhythm. A phase shifting is also possible by temperatures between 2° and 5°C (PAPI et al.).

The sand hopper *Talitrus saltator* is another example. It has been used in experiments of PAPI, PARDI, and PAPI, and of PARDI and GRASSI. The escape direction of this animal also depends on a sun compass and on a time sense. These animals usually live on the wet sand of the beach. If they are transferred to dry sand, they will escape back to the sea in a right angle to the shore line. This angle is determined on the basis of the sun's position. Some specimens of *Talitrus* were taken from Italy to Argentina. Their escape direction was oriented to the sun in accordance with their inner clock, which had been set in Italy. Furthermore, PARDI and GRASSI carried out experiments similar to the one described with starlings. They attempted to shift the clock by shifting the LD cycle. The result was as expected: the clock can be shifted (*PARDI).

The normal daily fluctuations of the environment do not influence the clock of *Talitrus*. Only very high temperature can cause a slight acceleration. Temperatures as low as 4° to 6°C did not impair the accuracy of the clock. This difference compared to bees and to *Arctosa* is not surprising. Apparently, the clock mechanism of several aquatic animals can only be disturbed by temperatures lower than those affecting other organisms; this has also been described for other circadian processes (Chapter 5(d)).

b. Peculiarities of certain species

Complete and incomplete compensation for the sun's movement. Some animals are still able to find the direction they had been trained for, even when they are exposed to an artificial sun during the night. In other words, they behave as though they knew that the movement of the sun continues below the horizon in the same direction as during the day; they compensate for a full 360°. This has been reported of starlings, bees, and fishes. According to this, starlings can make use of the sun compass during the entire arctic summer night in Lapland (HOFFMANN, 1959). However, this ability does not exist in all animals. In *Talitrus* (°PARDI), in the beetle *Phaleria provincialis* (PARDI 1958), the pond skater *Velia* (BIRUKOW u. BUSCH; EMEIS; °BIRUKOW), and also pigeons (SCHMIDT-KOENIG 1961), it has been found that these animals react during the night as though the sun were moving back in the opposite direction. The clock apparently runs in reverse direction during the dark periods of a 24-hour rhythm (Fig. 109).

> The ability to change the chosen direction in accordance with the course of time expires in *Velia* even within a few hours when the animal is kept without sunlight or in DD. In this case, the clock stops very quickly, so that the orientation rhythm does not continue endogenously (in contrast to most of the other objects which have been tested). But one should not deduce from this a basically different principle of time measurement, since we know that some other processes also controlled by the circadian clock are only loosely coupled to it.

We may conclude that complete or incomplete linking of the orientation to the clock depends very much upon the ecological re-

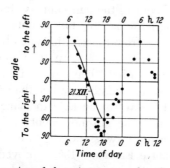

Fig. 109. *Velia currens* (pond skater). Mean angles of orientation of 10 animals in 10 successive trials to an artificial light source during daytime and at night. Ordinate: angle of orientation. Abscissa: Central European Time. Curve gives the sun's azimuth on Dec. 12. After BIRUKOW

quirements. PAPI and SYRJÄMÄKI published interesting results referring to the spider *Arctosa cinerea*. Animals from an Italian population are incapable of definite solar orientation during the night under the continuously visible sun of the arctic summer. They take up angles of orientation valid during the day or display a general picture of disorientation. Finnish populations of the same species, living within the arctic circle, orientate themselves correctly throughout the 24 hours of the days around the summer solstice.

Coupling with other diurnal functions. We have indicated the possibility that the different diurnal functions are dissociated. On the other hand, we also saw that several diurnal functions can be subject to a common control. This question has also been discussed in relation to the clock used in time estimation for sun-compass orientation.

FISCHER performed experiments with lizards (*Lacerta viridis*), testing the influence of changed light periods on the shifting of phases. The results indicate that the diurnal processes in light-compass orientation and in locomotive activity are controlled by the same inner clock. The pond skater *Velia currens* shows parallel rhythms in its light-compass orientation and its geotactic orientation (BIRUKOW u. OBERDORFER). According to the authors, this indicates a common central rhythm. Starlings, however, respond to a 6-hour phase shift of the LD cycle with an adaptation of their activity rhythm even after 2 days, whereas it takes 4 days to adapt the clock used in orientation (see RAWSON; *HOFFMANN 1960).

Adaptation to the direction of sun movement. The question arises as to how the organisms will adjust themselves to the different direction of sun movement in the northern and southern hemispheres. According to KALMUS, bees supposedly show hereditary differences. Investigations by LINDAUER, however, revealed that bees living in the tropical area can adjust themselves fairly quickly to the local direction of sun movement. Of course, the sun compass fails completely when the sun passes through the zenith, but orientation is possible even when the sun is only $2\frac{1}{2}°$ from the zenith. When bees were transferred in another experiment from a southern to a northern geographical latitude, they were able to adjust themselves within a few weeks to the sun's being in the south at noontime rather than in the north. Young bees need only about 8 days to be able to make the correct adjustments for the direction of the sun movement.

Fishes (*BRAEMER u. SCHWASSMANN), lizards (*BIRUKOW *et al.*) and crustaceans (PARDI e ERCOLINI 1966) proved to have the same learning ability. Thus, this seems to be a general rule.

Another phenomenon of learning. The possession of a clock does not necessarily mean that it is always used or that it always can be used. In physiological terms, this means that certain processes are coupled to the clock with various degrees of tenacity or perhaps not at all, depending on internal and external conditions. We have mentioned that some ants seem to account for the movement of the sun during orientation by sun compass, yet others do not. An experiment by JANDER seems to indicate that ants can learn to use their clock. During summer and autumn, individuals of *Formica rufa* were able to continue their way after a temporary halt by accounting for the changed position of the sun. In spring, they were unable to do so. It would, therefore, appear that ants first have to learn how to use their clock. Quite similar seasonal changes occur in coupling or uncoupling of sun orientation in the beetle *Pedaerus rubrothoracicus* (ERCOLINI e BADINO).

Apparently, there are other examples of this behavior. Lizards (*Lacerta viridis*) were trained for a certain angle to an artificial sun. If the experiment was always performed at the same time of day, they were able to maintain this angle later on. But they also chose this angle at another time of day; i.e., they did not take into account the movement of the sun. Only if different times of day are used during the training will the lizards choose the appropriate angle later on. Only then do they consider the light source as a moving sun, and not as a "landmark" (FISCHER u. BIRUKOW). Some fish also consider an artificial sun only as a "landmark," although most other individuals of the same species "evaluate" it as a sun, i.e., they choose different angles from it depending on the time of day (BRAEMER). Orientation by constant angle and by sun compass can compete with each other or replace each other in the pond skater *Velia currens* (EMEIS).

Considering seasonal changes. As mentioned above, most animals in using the sun compass consider only the azimuth but not the height of the sun. However, several authors have discussed the possibility that some animals are able to take into account both the height of the sun and the rate of azimuth change, varying with different times of day and year (see the discussions by *WALLRAFF; SCHWASSMANN and HASLER). FISCHER (1961) has reported a rather convincing example with lizards, where this ability is surprisingly well developed. Interestingly enough, increasing the day length under laboratory conditions results in a quick adaptation of these animals to sun-azimuth curves as they occur under natural conditions during longer days. Further details may be found in the following reviews: *BRAEMER u. SCHWASSMANN; *BIRUKOW et al.; *HASLER and SCHWASSMANN.

Orientation by the moon. The experimental evidence that the moon can also be used for orientation by light-compass is especially interesting. PAPI e PARDI (1953) investigated this in some detail especially in the previously mentioned *Talitrus saltator* (see reference to other organisms by °PAPI; PAPI and PARDI 1963; PARDI e ERCOLINI 1965). In this kind of orientation, the movement of the moon is correctly compensated for in time. This may permit the conclusion that these animals possess not only a physiological rhythm corresponding to the movement of the sun (i.e., the circadian rhythm), but also another rhythm corresponding to the movement of the moon. PAPI and PARDI published additional evidence in 1963 supporting the hypothesis that the lunar orientation of *Talitrus* is due to a continuously-operating lunar physiological rhythm. There are indications in support of this conclusion (see Chapter 11).

ENRIGHT has described the somewhat different response of the sand hopper *Orchestoidea corniculata*. At night these animals find their escape direction to the sea by moon orientation, but after being in the dark for a while they maintain a constant angle to the moon, without considering the time of day. It remains doubtful whether one can deduce from this a basically different mode of time measurement (see the remarks about the reactions of *Velia*, p. 171).

All in all, orientation by the moon is still somewhat controversial (for critical remarks, see °HOFFMANN 1965; ENRIGHT 1972).

References to Chapter 10

(*Use of the Clock in Direction Finding*)

a. Reviews

°AUTRUM, J., ed. 1963. *Animal Orientation: Symposium*. Ergebn. Biol. 26.

°BIRUKOW, G. 1960. Innate types of chronometry in insect orientation. In °CHOVNICK (ed.), pp. 403–412.—

°BIRUKOW, G., K. FISCHER, u. H. BÖTTCHER. 1963. Die Sonnenkompaßorientierung der Eidechsen. In °AUTRUM, pp. 216–234.—

°BRAEMER, W., u. H. O. SCHWASSMANN. 1963. Vom Rhythmus der Sonnenorientierung am Äquator (bei Fischen). In °AUTRUM, pp. 182–201.

°CHOVNICK, A., ed. 1960. *Biological Clocks*. Cold Spring Harbor Symp. quant. Biol. 25.

°GALLER, S. R., K. SCHMIDT-KOENIG, G. J. JACOBS, and R. E. BELLEVILLE, eds. 1972. *Animal Orientation and Navigation*. Washington, D.C.: Nat. Aeronautics and Space Administration.

°HASLER, A. D., and H. O. SCHWASSMANN. 1960. Sun orientation of fish at different latitude. In °CHOVNICK (ed.), pp. 429–441.—

°HOFFMANN, K. 1960. Experimental manipulation of the orientation clock in birds. In °CHOVNICK (ed.), pp. 379–387;—

———. 1965. Clock-mechanisms in celestial orientation of animals. In *Circadian Clocks.* J. Aschoff, ed. Amsterdam: North Holland Publishing Company. Pp. 426–441.
°Lindauer, M. 1960. Time compensated sun orientation in bees. In °Chovnick (ed.), pp. 371–377;–
———. 1963. Kompaßorientierung. In °Autrum, pp. 158–181.
°Matthews, G. V. T. 1968. *Bird Navigation.* 2nd ed. Cambridge: Univ. Press.
°Papi, F. 1960. Orientation by night: the moon. In °Chovnick (ed.), pp. 475–480.–
°Pardi, I. 1960. Innate components in the solar orientation of littoral amphipods. In °Chovnick (ed.), pp. 395–401.
°Schmidt-Koenig, K. 1960. Internal clocks and homing. In °Chovnick (ed.), pp. 389–393.–
°Schwassmann, H. O. 1960. Environmental cues in the orientation rhythm of fish. In °Chovnick (ed.), pp. 443–450.
°Wallraff, H. G. 1960. Does celestrial navigation exist in animals? In °Chovnick (ed.), pp. 451–461.

b. Other references

Birukow, G. 1956. *Z. Tierpsychol.* **13**:463–484.–
Birukow, F., u. E. Busch. 1957. *Z. Tierpsychol.* **14**:184–203.–
Birukow, G., u. H. Oberdorfer. 1959. *Z. Tierpsychol.* **16**:693–705.–
Braemer, W. 1960. *Verh. dtsch. zool. Ges. 1969.–*
Brun, R. 1958. *Abderhaldens Handbuch der biolog. Arbeitsmethoden.* Abteilung 6, S. 179; quoted from Renner.
Emeis, D. 1959. *Z. Tierpsychol.* **16**:129–154.–
Enright, J. T. 1961. *Biol. Bull.* **120**:148–156.–
———. 1972. In °Galler et al. (ed.), pp. 523–555.–
Ercolini, A., e G. Badino. 1961. *Bull. Zool.* **28**:421–432.
Ferguson, D. E., H. F. Landreth, and J. P. McKeown. 1967. *Anim. Behav.* **15**:45–53.–
Fischer, K. 1960. *Naturwiss.* **47**:287–288;–
———. 1961. *Z. Tierpsychol.* **18**:450–470;–
———. 1964. *Naturwiss.* **51**:203.–
Fischer, K., u. G. Birukow. 1960. *Naturwiss.* **47**:93.–
Frisch, K. v. 1950. *Experientia (Basel)* **6**:210–221.
Hasler, A. D. 1960. *Science* **132**:785–792.–
Hoffmann, K. 1954. *Z. Tierpsychol.* **11**:453–475;–
———. 1959. *Z. vergl. Physiol.* **41**:471–480.
Jander, R. 1957. *Z. vergl. Physiol.* **40**:162–238.
Kalmus, H. 1956. *J. exp. Biol.* **33**:554–565.–
Keeton, W. T. 1969. *Science* **165**:922.–
Kramer, G. 1953. *Verh. dtsch. zool. Ges. Freiburg 1952,* S. 78–84. Leipzig.
Lindauer, M. 1957. *Naturwiss.* **44**:1–6;–
———. *Z. vergl. Physiol.* **42**:43–62.–

LÜTERS, W., u. G. BIRUKOW. 1963. *Naturwiss.* **50**:737–738.
MICHENER, M. C., and CH. WALCOTT. 1966. *Science* **154**:410–413.–
MISELIS, R., and CH. WALCOTT. 1970. *Anim. Behav.* **18**:544–551.
PAPI, F. 1955. *Z. vergl. Physiol.* **37**:230–233.–
PAPI, F., L. FIORE, V. FIASCHI, and N. E. BALDACCINI. 1971. *Z. vergl. Physiol.* **73**:317–338.–
PAPI, F., e. L. PARDI. 1953. *Z. vergl Physiol.* **35**:490–518.–
PAPI, F., and L. PARDI. 1963. *Biol. Bull.* **124**:97–105;–
———. 1968. *Monitore Zool. Ital.* N.S. **2**:217–231.–
PAPI, F., and J. SYRJÄMÄKI. 1963. *Arch. ital. Biol.* **101**:59–77.–
PAPI, F., L. SERRETTI, e S. PARRINI. 1957. *Z. vergl. Physiol.* **39**:531–561.–
PARDI, L. 1958. *Atti Acad. Sc. Torino* **92**:1–8;–
———. 1958. *Z. Tierpsychol.* **14**:261–275.–
PARDI, L., e. A. ERCOLINI. 1965. *Z. vergl. Physiol.* **50**:225–249;–
———. 1966. *Monitore Zool. Ital.* **74**: Suppl. 80–101.–
PARDI, L., e M. GRASSI. 1955. *Experientia (Basel)* **11**:202–210.–
PARDI, L., u. F. PAPI, 1952. *Naturwiss.* **39**:262–263;–
———. 1953. *Z. vergl. Physiol.* **35**:459–489.
RAWSON, K. S. 1956. Ph.D. thesis. Harvard Univ.
SCHMIDT-KOENIG, K. 1958. *Naturwiss.* **45**:47;–
———. 1958. *Z. Tierpsychol.* **15**:301–331;–
———. 1961. *Naturwiss.* **48**:110;–
———. 1961. *Z. Tierpsychol.* **18**:221–244;–
———. 1963. *Biol. Bull.* **124**:311–321;–
———. 1964. *Biol. Bull.* **127**:154–158;–
———. 1969. *Zool. Anz. Suppl.* **33**:200–205;–
———. 1970. *Z. vergl. Physiol.* **68**:39–48.–
SCHMIDT-KOENIG, K., and D. L. MCDONALD. 1970. *Science* **168**:152–153.–
SCHWASSMANN, H. O., and W. BRAEMER. 1961. *Physiol. Zool.* **34**:273–286.–
SCHWASSMANN, H. O., and A. D. HASLER. 1964. *Physiol. Zool.* **37**:163–178.

11. Relations between Circadian, Tidal, and Lunar Rhythms

> Inquiries for Surratte, and other parts of the East Indies . . . Whether those shell-fishes, that are in these parts plump and in season at the full moon, and lean and out of season in the new, are found to have contrary constitutions in the East Indies?
>
> Philos. Transact. Royal Soc. London 2, 419 (1666).

a. Endogenous tidal rhythm

Examples. Numerous marine organisms show rhythms in their behavior, coinciding with the cycles of high and low tide that continue under laboratory conditions. BOHN (1903) found that green flatworms (*Convoluta*) come to the surface of the sand during high tide and that at low tide they bury themselves in the sand as it dries. This rhythm also continues in an aquarium without the tidal cycle. We may further refer to the investigations by BRACHER in *Euglena limosa,* by FAURÉ-FREMIET in *Chromulina* and in the diatom *Hantzschia.*

Several authors have described the continuations of these tidal rhythms under laboratory conditions in mussels and Crustaceae as well (see, e.g., BENNETT; BROWN *et al.;* RAO; FINGERMAN; CHANDRASHEKARAN; MORGAN, and also Figs. 110, 111). BOHN (1906) has reported that the sea anemone *Actinia equina,* kept in an aquarium, continues to expand and contract for up to 8 days.

In several cases earlier observations on the persistence of the tidal cycles have been doubted (for example, with respect to *Actinia* by DI MILA and GEPPETTI 1964a and b). But it is of some importance that the relative strength of tidal and circadian rhythms is very much influenced by differences of the habitat as demonstrated by BARNWELL with several *Uca* species (fiddler crab). There are even striking individual differences in species of the same habitat (Fig. 110).

Nevertheless, there are species that show the phenomenon of persistent tidal rhythms very clearly. One of the best examples is the littoral fish *Blennius pholis,* which shows a persistence of cycles in swimming activity for at least 5 days with periods of 12.56 hours in

178   The Physiological Clock

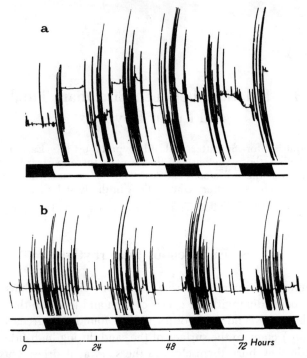

Fig. 110. Running activity of the shore crab *Carcinus maenas* in LD cycles. Horizontal strips indicate conditions of illumination (a) approximately 12-hour periodicity, (b) 24-hour periodicity. Original

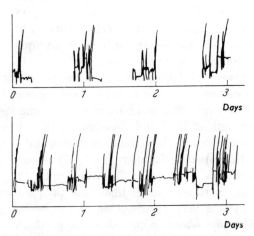

Fig. 111. *Carcinus maenas*. As Fig. 110, but in LL. Original

DD, 12.5 hours in LL (GIBSON 1965). Thus, the periods of the free-running rhythm are a little longer than the normal ones of 12.4 hours. The sand beach isopod *Eurydice pulchra* (Crustacea) has an endogenously controlled tidal swimming rhythm (JONES and NAYLOR). The arctic population of the intertidal midge *Clunio marinus* shows a tidal periodicity of emergence. Emergence occurs between the times of high and low tide (NEUMANN and HONEGGER).

Interaction with circadian rhythms. In other cases, the tidal rhythm can be interacting with a circadian rhythm, or the circadian rhythm even replaces the tidal one. These genetic differences are mainly adaptations to the special ecological conditions. In the case of *Clunio* mentioned above, the emergence rhythm of populations from temperate zones is characterized by a combination of a semilunar and a diurnal periodicity (NEUMANN 1968, *1969).

NAYLOR (1958) studied the interaction of a circadian rhythm with a circatidal rhythm in the European shore crab *Carcinus maenas*.

Fig. 110 gives two examples of the activity rhythm during diurnal LD cycles. One can see that either the 24-hour component dominates with a more intense locomotive activity during the dark periods, or it is superimposed upon a secondary periodicity. The mathematical analysis of these curves reveals a periodicity of about 12.4 hours besides the normal 24-hour periodicity. This behavior continues in LL (weak light), sometimes demonstrating only the 24-hour component (Fig. 111; see BLUME et al.). A similar interaction was also reported for another intertidal crab (*Sesarma reticulatum*, see PALMER) and for the fish *Coryphoblennius galerita* (GIBSON 1971).

It is not surprising that individual differences exist in the relative intensity of the two components. The differences are even more striking when shore crabs from different localities are compared. In *Carcinus mediterraneus* from Naples, NAYLOR (1961) found a pronounced 24-hour rhythm in LL (dim light). Again, the timing of the maxima coincided with the dark period preceding the experiment but the tidal rhythm was hardly detectable. In general, the animals responded as in Fig. 110. Only a few showed indications of the tidal rhythm. These observations can be correlated to the ecological conditions since the tidal change in the Bay of Naples is insignificant.

In all these examples, one cannot exclude the possibility that the rhythmicity with about 12.4-hour periods results from a superposition of two approximately 24.8-hour periods. Since the tidal rhythmicity is synchronized by cues other than the LD cycles, this would mean a superposition of two circadian rhythms that are not predominantly coupled to light receptors. Experiments with the sand beach isopod *Excirolana chiltoni* support this interpretation: two oscillators seem to operate simultaneously in each individual, the period of each oscillator being about 24 hours (KLAPOW,).

The interaction of two periods of about 24.8 hours with a 24-hour rhythmicity would mean a superposition of two circadian rhythms unaffected by light cues with the "normal" light-controlled circadian rhythm. Observations concerning more complicated cases support this interpretation.

More complicated cases. Adaptations to irregular changes of high and low tide are possible. Such conditions are found, for example, on the California coast. It is remarkable that a corresponding behavior also continues under laboratory conditions: a longer and a shorter distance between two peaks may alternate (ENRIGHT 1963). The length of the free-running periods that are no longer controlled by the tides can deviate from the length of synchronized periods. They may be longer, for example, as in diurnal rhythms that are no longer synchronized by external factors.

These results indicate that we are not dealing with a rhythm of periods of about 12 to 14 hours, but with the superposition of approximately 24-hour rhythms. The findings of ENRIGHT (1963), working with *Synchelidium sp.* (an intertidal amphipod), support this interpretation. This animal from the California coast behaves in a way similar to that mentioned (Fig. 112) and continues to do so for several days in the laboratory. Apparently, rhythms with a free-running period of about 26 hours under laboratory conditions (i.e., without any synchronizers) are responsible for this.

Still more remarkable is the behavior of another intertidal isopod, *Excirolana chiltoni,* from the California coast (ENRIGHT 1972). Under constant conditions, this "virtuoso isopod" (ENRIGHT) showed a persistent rhythm in its swimming activity. The free-running period was about 24.9 hours. During the two months of observation the amount of activity per burst showed a periodic amplitude modulation that paralleled the complex lunar cycle of change in high and low tide, characteristic for the California coast. The free-running period of the circa-lunar rhythm of amplitude modulation was 1 or 2 days longer than the natural 29-day lunar cycle of tide heights.

> . . . it seems clear that the endogenous tidal rhythm of *Excirolana,* with its circa-lunar amplitude modulation, is able to provide a much more complex recapitulation of preceding environmental conditions than are those circadian rhythms previously described. An animal with a circadian rhythm can, in a way, be compared with a drummer, often capable of considerable precision in keeping time when isolated from the orchestra of natural day-night cycles. To carry on with that analogy, *Excirolana* is capable not only of maintaining tempo, but of repeating entire ornamented phrases of an environ-

mental score. Even within the framework of this accomplishment of the species, the behavior is a virtuoso performance—the animal's left hand playing the melody, the right hand the counterpoint (ENRIGHT 1972).

Similarities with the circadian rhythm. A close relationship of the tidal rhythm with the circadian rhythm can also be deduced from the fact that the endogenous tidal rhythm is to a large extent independent of temperature (see RAO; NAYLOR 1963). Even more striking is the fact that shore crabs that show only the circadian rhythmicity and not the tidal component (p. 179) can be induced to reveal the tidal rhythmicity. A single period of chilling to 4° (for 15 hours) is sufficient for this transformation (WILLIAMS and NAYLOR). An additional argument might be the fact that D_2O lengthens the period of the endogenous tidal rhythm (ENRIGHT 1971), comparable with the D_2O-effect on circadian rhythms (p. 140).

Should further experiments confirm the assumption that tidal rhythms result from the superposition of two nearly 24.8-hour rhythms (p. 179), the main difference between these and other "normal" cir-

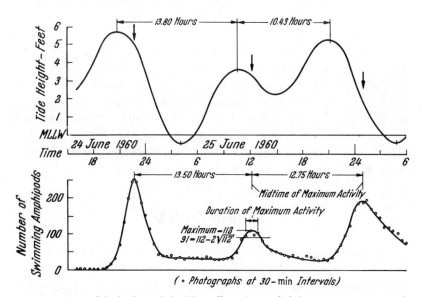

Fig. 112. Tidal rhythm of freshly collected *Synchelidium* sp.; upper graph: tide height, from U.S. Coast and Geodetic Survey predictions for La Jolla, California; MLLW: zero tide reference level = mean lower low water; lower graph: counts of number of swimming amphipods during laboratory observation; two photographs per hour, Center lines of activity peaks determined as midtime of maximum activity, with calculations indicated on second activity peak. Vertical arrows over tide profile correspond to center lines of activity peaks. After ENRIGHT 1963

cadian rhythms would mainly be in the effective synchronizers. Tidal rhythms are not predominantly light-controlled.

A further argument for seeking the main difference between "normal" circadian and tidal rhythm in the effective synchronizers is the behavior of the intertidal prawn *Palaemon elegans*. Its rhythm quickly rephases to LD cycles during neap tides, but shows again the features of other tidal rhythms when spring tides return, i.e., synchronization by cues connected with the fall and rise of water (RODRIGUEZ and NAYLOR).

The activity cycles in the fiddler crab *Uca crenulata* can also be synchronized by LD cycles. In this case, the responsiveness to light can only be observed in animals investigated in April and July (HONEGGER).

Thus, the tidal rhythmicity has features in common with the phenomenon of splitting described for some cases of circadian rhythmicity (p. 64, Chapter 14, and Fig. 133; see also the discussion in Chapter 6(b)).

Cues. It is likely that the different organisms respond quite differently to external factors. With fluctuating water levels the organisms are exposed to a change of high and low light intensity, but the fluctuating hydrostatic pressures might also be important (MORGAN; GIBSON 1971). ENRIGHT (1961) found that an amphipod from the tidal zone can react to pressure changes of less than 0.01 atmospheres. In 1963, ENRIGHT observed that the *Synchelidium* species from the California coast responds primarily to the turbulence of streaming water as the decisive synchronizer in tidal changes. This holds also for the sand beach isopod *Excirolana chiltoni* (KLAPOW). Variations in hydrostatic pressure and wave action are synchronizers for the sand beach isopod *Eurydice pulchra* (JONES and NAYLOR).

b. Lunar rhythms

General observations and examples. Lunar cycles have been known in several marine organisms for some time. Reproduction in some of them, for example, is restricted to the time of the spring or neap tides, or to days that bear some fixed relation to these times. The physiological rhythms that then occur correspond to the lunar cycle of 28.5 days or in some other cases to one half this cycle, i.e., to about 14 to 15 days (*CASPERS; *HAUENSCHILD; *CLOUDSLEY-THOMPSON).

These phenomena are so different, at first glance, from the phenomena of diurnal rhythms that they seem to be out of place in this context. However, some relations to the endogenous diurnal rhythm have become evident.

Some of the better known examples will be mentioned first. The palolo-worm of the Pacific and the Atlantic is probably most famous of all. This animal reproduces only twice a year, namely, during the neap tides of the last quarter moon in October and November. The grunion fish (*Leuresthes tenuis*) of the California coast takes advantage of the highest tides. Riding on the crest of the waves, it arrives on the beach, where it deposits its eggs and sperm. The fertilized eggs develop in the warm moist sand, and water does not reach them for the next two weeks. Only by the next spring tide have the fishes developed sufficiently to be freed from their eggs and washed out into the open sea. Activities with such a lunar periodicity have also been described with mussels (e.g., oysters, LOOSANOFF and NOMEJKO; KNIGHT-JONES), for sea urchins and for several other animal and plant organisms of the sea (see also CHRISTIE and EVANS on the discharge of gametes by the alga *Enteromorpha intestinalis*). The brown alga *Dictyota dichotoma* liberates its male and female gametes at certain places twice within a lunar cycle, i.e., 14 to 15 days apart (HOYT).

But lunar cycles occur also in species that do not live in water. In the beetle *Calandra granaria*, BIRUKOW observed cycles in phototactic responsiveness that corresponded to the lunar phases. Numbers of the moth *Heliothis zea* (bollworm) caught in light traps showed rhythmic increases and decreases corresponding to the lunar phases (NEMEC). There is also experimental evidence for the occurrence of lunar cycles in mammals, even in primates.

A remarkable feature of these processes is that several species continue the periodicity under laboratory conditions, i.e., without being exposed to tidal changes or to moon light. This very strongly suggests the existence of an endogenous lunar rhythm. STRUMWASSER was even able to demonstrate a persisting fornightly lunar rhythm in one of the nerve cells of an isolated gaglion from the sea hare *Aplysia californica*.

Synchronization by the moon. Moonlight can control the phases of this endogenous lunar rhythm. HAUENSCHILD has demonstrated this for the marine worm *Platynereis*, where the phases can be determined in the laboratory by artificial moonlight, i.e., illumination at night in the range of moonlight intensity. This also holds true for the brown alga *Dictyota*, which exhibits these relations even more clearly (Figs. 113, 114). The maximum discharge of eggs takes place 9 days after exposure to moonlight. The next maximum then follows after an interval of 15–16 days. It should not be surprising that this free-running period is somewhat longer than half the lunar cycle, since we have

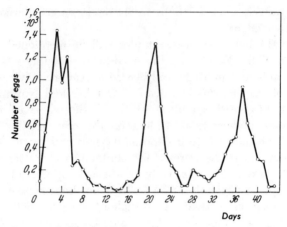

Fig. 113. *Dictyota dichotoma*. Release of eggs into the water. Plants growing in LD 14:10. Peaks 16 days apart. After BÜNNING u. MÜLLER

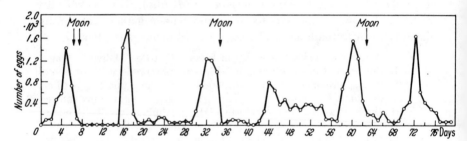

Fig. 114. Similar to Fig. 113, but in this case "artificial moonlight" (light during dark periods) was offered at times indicated by arrows. Peaks occur 9 days after moon. Distance of peaks without moon again 16 days as in Fig. 113. After BÜNNING u. MÜLLER

also found similar differences in other endogenous rhythms, when synchronizing factors are missing. In certain populations of *Clunio marinus* moonlight has the same controlling influence (NEUMANN 1965, 1966). In general we may say that lunar cycles of about 29 or 15 days are endogenous in the more closely investigated cases and that the phases may be determined by moonlight.

Other synchronizers. The diurnal LD cycle in combination with a tidal factor may also be active in synchronizing the lunar cycle. This holds for a certain population of *Clunio* (NEUMANN 1968). Concerning *Excirolana*, ENRIGHT (1972) supposes that some factor linked with tidal height acts as synchronizer.

Interpretation based upon two interacting rhythms. Several workers have proposed that these lunar or semilunar cycles find their explanation in the interaction of diurnal rhythms with tidal rhythms. Small differences in the period length of the two rhythms would result in a coincidence of certain phases, i.e., a reinforcement or beat reaction occurring at intervals of about 15 days or about 29 days. BROWN and coworkers have indicated this possibility for the fiddler crab *Uca*. Studying the rhythm of melanophore movement, they found a component with a period of 24 hours and another one with a period of 12.4 hours. Furthermore, the experiments of NAYLOR with the shore crab *Carcinus maenas* should again be noted here (see p. 179). Two components could be distinguished in the locomotive activity cycle. It is possible to observe a shift of the peaks of two rhythms, increasing from day to day according to the difference between the tidal cycle and the diurnal cycle. This continues in the laboratory without the influence of tidal changes. The result of this shift is a coincidence of the two maxima at intervals of about 15 days.

There should be yet another way to test the correctness of this interpretation. The *circadian* rhythm can be controlled to a great extent by cycles of light and darkness. It should be possible to entrain the period not only to 24 hours but also to somewhat shorter or longer cycles. If the light-dark cycle does not influence the period of the *tidal* rhythm, the coincidence in the maxima of the two rhythms should occur sooner—when the difference between the periods has been increased and later when it has been decreased.

In *Dictyota* this does seem to hold true. This brown alga was exposed to "days" (light plus darkness) of more or less than 24 hours. Under these conditions the maximum discharge of eggs from the oogonia did not take place with intervals of about 15 days, but with intervals as shown in Table 7 (BÜNNING u. MÜLLER; MÜLLER; VIELHABEN).

The influences are in the expected direction, but they are less than should be expected if the tidal cycles of 12.4 hours would continue as free-running rhythms without being influenced by the LD cycles. Any

Table 7. *Dictyota dichotoma*.
Influence of "False" Days on the Lunar Cycle

Length of the Entraining LD Cycle	Length of the Period in Egg Discharge
hrs	days
23.0 (L:D = 13.5:9.5)	11–12
23.5 (L:D = 13.75:9.75)	12–13
24.0 (L:D = 14.0:10.0)	14–16
24.5 (L:D = 14.25:10.25)	16–17

final conclusions on the "beat hypothesis" of lunar cycles would require detailed studies on the running of the circadian and tidal rhythms under conditions like those mentioned.

*HAUENSCHILD (1960) published results from experiments with *Platynereis* which cannot be interpreted as mentioned above. Similarly, experiments with the green alga *Halicystis parvula* (ZIEGLER PAGE and SWEENEY) and with the isopod *Excirolana chiltoni* (ENRIGHT 1972) do not support the beat hypothesis of lunar and semilunar cycles.

Ecological importance. In most cases the ecological importance of lunar cyclical phenomena is quite obvious. Reproduction can be adjusted to the environmental conditions of spring tide and neap tide. In some cases synchronization in the release of male and female gametes is even more important. The chances of fertilization are considerably increased. In addition to restricting reproduction to a few days in the lunar cycle, there is often also a restriction to a certain time of year and even to a certain time of day. By such means the reproductive processes in marine organisms, both plants and animals, can sometimes be restricted to a few hours in an entire year. This holds, for example, for the alga *Dictyota* (VIELHABEN) and perhaps even more strictly for some marine animals (see KORRINGA). By this means the chances of male and female gametes uniting would be increased a thousandfold over the chances if liberation were to be at random.

In all cases, the endogenous nature of the rhythmicity enables the organism to find the "appropriate" time in nature, even when the moon is hidden for 2 or 3 months because of an overcast sky.

References to Chapter 11

(Interaction of Circadian Cycles with Tidal and Lunar Cycles)

a. Reviews

*CASPERS, H. 1951. Rhythmische Erscheinungen in der Fortpflanzung von *Clunio marinus* (Dipt. Chiron.) und das Problem der lunaren Periodicität bei Organismen. *Arch. Hydrobiol. Suppl.* 18:415–494.–

*CLOUDSLEY-THOMPSON, J. L. 1961. *Rhythmic Activity in Animal Physiology and Behaviour*. New York–London: Academic Press.

*FINGERMAN, M. 1960. Tidal rhythmicity in marine organisms. *Cold Spring Harbor Symp. quant. Biol.* 25:481–489.

*HAUENSCHILD, C. 1960. Lunar periodicity. *Cold Spring Harbor Symp. quant. Biol.* 25:491–497.

°NEUMANN, D. 1969. Die Kombination verschiedener endogener Rhythmen bei der zeitlichen Programmierung von Entwicklung und Verhalten. *Oecologia* 3:166–183.

b. Other references

BARNWELL, F. H. 1963. *Biol. Bull.* 125:399–415.–
BENNETT, M. F. 1954. *Biol. Bull. Woods Hole* 107:174–191.–
BIRUKOW, G. 1964. *Z. Tierpsychol.* 21:279–301.–
BLUME, J., E. BÜNNING, u. D. MÜLLER. 1962. *Biol. Zentralbl.* 81:569–573.–
BOHN, G. 1903. *C. R. Acad. Sci. (Paris)* 37:576–578;–
———. 1906. *C. R. Acad. Sci. (Paris)* 61:420–422.–
BRACHER, R. 1954. *J. Limn. Soc. Physiol.* 44:477–505.–
BROWN, F. A., M. FINGERMAN, M. I. SANDEEN, and H. M. WEBB. 1953. *J. exp. Zool.* 123:29–30.–
BROWN, F. A., M. FINGERMAN, and M. N. HINES. 1954. *Biol. Bull.* 106:308–317.–
BÜNNING, E., u. D. MÜLLER. 1962. *Z. Naturforsch.* 16b:391–395.
CASPERS, H. 1961. *Int. Rev. ges. Hydrobiol.* 46:175–183.–
CHANDRASHEKARAN, M. K. 1965. *Z. vergl. Physiol.* 50:137–150.–
CHRISTIE, A. O., and L. V. EVANS. 1962. *Nature* 193:193–194.–
CORBET, P. S. 1960. *Cold Spring Harbor Symp. quant. Biol.* 25:357–360.
DI MILA, A., and L. GEPPETTI. 1964a. *Experientia* 20:571–572.–
DI MILA, A., e. L. GEPPETTI. 1964b. *Monitore Zoolog. Italiano* 72:203–214
ENRIGHT, J. T. 1960. *Cold Spring Harbor Symp. quant. Biol.* 25:487–489;–
———. 1961. *Science* 133:758–760;–
———. 1963. *Z. vergl. Physiol.* 46:276–313;–
———. 1971. *Z. vergl. Physiol.* 72:1–16;–
———. 1972. *J. comp. Physiol.* 77:141–162.
FAURÉ-FREMIET, E. 1950. *Bull. biol. France et Belg.* 84:207–214.–
FINGERMAN, M. 1955. *Biol. Bull. Woods Hole* 109:255–264.
GIBSON, R. N. 1965. *Nature* 207:544–545;–
———. 1967. *J. mar. biol. Ass. U. K.* 47:97–111.
———. 1971. *Anim. Behav.* 19:336–343.
HARTLAND-ROWE, R. 1958. *Rev. zool. botan. africaine* 58:185–210.–
HAUENSCHILD, C. 1955. *Z. Naturforsch.* 10b:658–662.–
HONEGGER, H.-W. 1973. *Marine Biol.* 18:19–31.
HOYT, W. D. 1927. *Amer. J. Bot.* 14:592–619.
JONES, D. A., and E. NAYLOR. 1970. *J. exp. mar. Biol. Ecol.* 4:188–199.
KLAPOW, L. A. 1972. *J. comp. Physiol.* 79:233–258.–
KNIGHT-JONES, E. W. 1952. *Fishery Investigation Ser. II,* 18(2):1–48.–
KORRINGA, P. 1947. *Ecol. Monogr.* 17:347–348.
LANG, H. J. 1967. *Z. vergl. Physiol.* 56:296–340.–
LOOSANOFF, V. L., and C. A. NOMEJKO. 1951. *Ecology* 32:113–134.
MORGAN, E. 1965. *J. anim. Ecol.* 34:731–746.–
MÜLLER, D. 1962. *Botan. Marina* 4:140–155.

NAYLOR, E. 1958. *J. exp. Biol.* **35**:602–610;–
———. 1960. *J. exp. Biol.* **37**:481–488;–
———. 1961. *Pubbl. Staz. Zool. Napoli* **32**:58–63;–
———. 1963. *J. exp. Biol.* **40**:669–679.–
NEMEC, S. J. 1971. *J. Econ. Entomol.* **64**:860–864.–
NEUMANN, D. 1965. *Z. Naturforsch.* **20b**:818–819;–
———. 1966. *Z. vergl. Physiol.* **53**:1–61;
———. 1968. **60**:63–78.–
NEUMANN, D., and H. W. HONEGGER. 1969. *Oecologia* **3**:1–13.
PALMER, J. D. 1967. *Nature* **215**:64–66.
RAO, K. P. 1954. *Biol. Bull. Woods Hole* **106**:353–359.–
RODRIGUEZ, G., and E. NAYLOR. 1972. *J. mar. biol. Ass. U. K.* **52**:81–95.
STRUMWASSER, F. 1965. In *Circadian Clocks*. J. ASCHOFF, ed. Amsterdam: North Holland Publishing Company. Pp. 442–446.
VIELHABEN, V. 1963. *Z. Bot.* **51**:156–173.
WILLIAMS, B. G., and E. NAYLOR. 1967. *J. exp. Biol.* **47**:229–234.
ZIEGLER PAGE, J. Z., and B. M. SWEENEY. 1968. *J. Phycology* **4**:253–260.

12. Control of Diurnal Fluctuations in Responsiveness to External Factors

a. General remarks

The sensitivity to external factors can be influenced by the circadian clock; i.e., the organism responds differently to stimuli of equal intensity depending on the phase of the endogenous rhythm. Quantitative or even qualitative differences may occur.

> For some time phenomena have been known, especially in humans, that may indicate that the sensitivity of the organism to external factors does not fluctuate at random (references: KAISER and HALBERG). The extent to which these fluctuations are controlled by the endogenous clock is not known. Most reports on the subject refer to individuals not living in constant conditions during the experiments.

We shall restrict our discussion here to cases in which it is likely that the diurnal fluctuations in sensitivity are controlled by the circadian clock, i.e., cases in which these fluctuations continue for some time in LL or DD, corresponding to other endogenous diurnal phenomena.

b. Responsiveness to light

We are already familiar with one example in which the circadian rhythm controls fluctuations in the responsiveness to light: the circadian rhythm of phototactic responsiveness in *Euglena*. Another circadian rhythm in a photoresponse of a unicellular alga was reported by FORWARD and DAVENPORT. Even qualitative changes in the responsiveness to light may occur. The unicellular alga *Micrasterias denticulata* shows a circadian rhythm in the ratio of efficiencies of red and blue light as factors for speed of motion and ability of orientation (NEUSCHELER).

Daphnia also shows periodic fluctuations in responsiveness to light which continue in DD (RIMET). Among insects, some prefer different intensities of light depending on the time of day (REMMERT). Some species avoid light at night but seek it during the day. Others respond in the opposite way. It is also interesting that the electrographs from the compound eyes of beetles reveal diurnal fluctuations. These electric responses after illumination show differences not only in quantity but also in quality, i.e., in the shape of the curves of these electrographs (JAHN and CRESCITELLI).

The isolated eyes of the sea hare *Aplysia* show circadian rhythms of sensitivity to light pulses, as measured by the frequency of impulses evoked by illumination. These rhythms persist in DD (JACKLET).

> For additional examples of quantitative or qualitative diurnal changes in the responsiveness of animals to light, see BIRUKOW (in the beetle *Calandra granaria*) and PALMER (in *Uca pugnax*). In several cases these cycles are known to persist under constant conditions.

We have already mentioned the diurnal fluctuations in the ability to synthesize chlorophyll and the fluctuations in photosynthetic capacity. The continuation of these phenomena under constant conditions has been clearly demonstrated.

Some of the phenomena cited demonstrate that different reactions— not only quantitatively but also qualitatively—may occur after stimuli of equal intensity given at different times of the day, or under different physiological conditions as controlled by the endogenous clock. Qualitative fluctuations in responsiveness to light become evident by testing the development of plants and animals after they are exposed to light stimuli of equal intensity applied during different phases of the rhythm. These fluctuations are the basis of the so-called photoperiodic reactions. We shall discuss this in Chapter 13, which deals with day-length measurement.

c. Responsiveness to temperature

Quite often the circadian clock also controls the temperature sensitivity of organisms. The fluctuations in sensitivity resulting from this may be expressed, for example, in diurnal fluctuations of cold and heat resistance and in diurnal fluctuations of the optimum temperature for development.

Fluctuating resistance. The investigations by SCHWEMMLE; SCHWEMMLE u. LANGE of the plant *Kalanchoe* offer an example of fluctuating resistance. The extent of damage on leaves caused by high temperatures varies with the phase of the endogenous diurnal rhythm (Fig. 115).

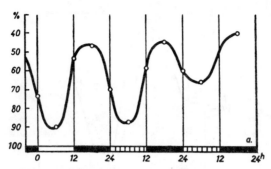

Fig. 115. Endogenous daily fluctuations of heat resistance of the leaves of *Kalanchoe blossfeldiana*. 12:12 hour LD cycles, followed by DD. Times that would have been light periods (during the 60 hours of DD) are indicated by the vertically striped bars (between 24 and 12). Ordinate: damage (percent of leaf area). After SCHWEMMLE u. LANGE

Varying influence on development. Plants also show sensitivity fluctuations when the applied temperatures are not extreme. High or low temperatures influence growth and other developmental processes quite differently, perhaps even with a qualitative difference, depending on the phase of endogenous rhythm. The temperature optimum of a certain developmental process also changes with the phases of the endogenous rhythm. For optimum plant development, low temperatures are required when the endogenous rhythm corresponds to the dark period and high temperatures are required during the opposite phases. These sensitivity fluctuations or diurnal changes in temperature optimum are the basis of the so-called thermoperiodic reactions.

Animals can also show this preference of diurnal temperature cycles, as HEATH has shown in *Salmo clarki clarki*.

d. Susceptibility to other factors

Chemical factors. Examples of how sensitivity to poisons, drugs, and many other factors may fluctuate diurnally have been described repeatedly (Fig. 116; see also HALBERG et al.; HAUS and HALBERG). Yet no one has checked to see if *all* these diurnal fluctuations of the organism also continue under *constant* conditions if equal doses of poison, etc., are applied. Therefore, no general statements can be made as to whether these fluctuations are controlled only by the endogenous rhythm, or whether external factors might also have a direct influence.

Further examples of diurnal changes in responsiveness to drugs, poisons, etc., have been published by DAVIS and WEBB; EMLEN and

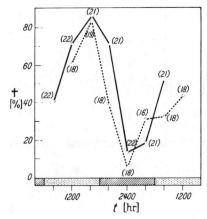

Fig. 116. Physiologic 24-hour periodicity and the endpoint "death from an endotoxin." Ordinate: $+\% = \%$ death of group of standardized mature C mice from *E. coli* lipopolysaccharide (Difco, 100 µg/20 g i.p.). Abscissa: time of injection, in two experiments (injections begun at two different time points, during daily light period). Number of mice/group in parenthesis. After HALBERG et al.

KEM; ERTEL et al.; HALBERG; MOTTRAM; POLCIK et al.; LUTSCH and MORRIS; SCHEVING et al.; *REINBERG. Diurnal changes in radiosensitivity have been observed in plants (HOFER u. BIEBL; BIEBL u. HOFER); in animals they are rather small or not significant (PIZZARELLO et al.; SPALDING and McWILLIAMS; LINDBERG et al.). In several cases these cycles of sensibility are connected with mitotic cycles.

Fluctuations in susceptibility to drugs are eminently important in connection with the application of drugs. This was pointed out by *JORES (1938). He suggested, for example, that bile-promoting drugs should not be administered in the evening, because that is when the liver is already in the phase of glycogen formation (see also Fig. 12). Another example of interest, since we have recognized mitosis as a process frequently controlled by the endogenous rhythm, is the EHRLICH-ascites-carcinoma in mice, which also follows a diurnal rhythm in its mitosis (MENG u. POHLE). This rhythm apparently accounts for the diurnal fluctuation in the effect of mustard gas on this carcinoma. The poison is less effective when applied just before maximum mitotic activity than when applied before minimum mitotic activity (POHLE et al.).

This observation may be correlated with the molecular biology of the cell division cycle. The period of DNA synthesis ends some time before mitosis commences (STANNERS and TILL). Since nitrogen mustard acts on DNA and eventually prevents DNA synthesis it would be expected to be more effective in the period after mitosis than just before mitosis.

Other examples, such as the application of insulin or the effect of drugs affecting the heart or circulation, appear in °MENNINGER-LERCHENTHAL and by MENZEL.

°HALBERG emphasizes the importance from the medical viewpoint of realizing that the endogenous rhythm created *horae minoris resistentia* and *horae variae resistentiae*, respectively.

Very impressive in this context are observations on diurnal variations of responses to prolactin in several lower vertebrates. The responses were even antagonistic (causing fattening versus a loss in fat stores) in some cases. "The diurnal variations in responses to prolactin offer an explanation for the conflicting and diverse reports regarding the function of prolactin and emphasize the importance of time in understanding the physiology of prolactin" (MEIER).

BURNS *et al.* described another example of circadian rhythmicity in the responses to a substance with antagonistic effects. The uptake of labeled thymidine by the mouse parotid gland, kidney, and duodenum follows a circadian pattern. Isoprotorenol may either stimulate or inhibit this uptake, depending on the phase of application.

X-irradiation. Several authors have reported on circadian rhythms in sensitivity to X-rays (for examples and references, see RENSING).

References to Chapter 12

(Control of Diurnal Fluctuations in Responsiveness to External Factors)

a. Reviews

°HALBERG, F. 1960. The 24-hour scale: a time dimension of adaptive functional organization. *Persp. Biol. Med.* 3:491–527;—
———. 1962. Physiological 24-hours rhythms: a determinant of response to environmental agents. In *Man's Dependence on the Earthly Atmosphere.* K. E. SCHAEFER, ed. New York: Macmillan. Pp. 48–99.—
°HAUS, E. 1964. Periodicity in response and susceptibility to environmental stimuli. *Ann. N. Y. Acad. Sci.* 117:292–315.
°JORES, A. 1937. Die 24-Stunden Periodik in der Biologie. *Tabulae biologicae (Den Haag)* 14:77–109;—
———. 1938. Endokrines und vegetatives System in ihrer Bedeutung für die Tagesperiodik. *Dtsch. med. Wschr.* Nr. 21 u. 28.
°MENNINGER-LERCHENTHAL, E. 1960. Periodizität in der Psychopathologie. Wien: Maudrich-Verl.

°REINBERG, A. 1965. Hours of changing responsiveness in relation to allergy and the circadian adrenal cycle. In *Circadian Clocks*. J. ASCHOFF, ed. Amsterdam: North Holland Publishing Company. Pp. 214–218;—
———. 1967. The hours of changing responsiveness or susceptibility. *Perspect. Biol. Med.* 11:111–128.

b. Other references

BIEBL, R., u. K. HOFER. 1960. *Radiation Biol.* 6:225–250.—
BIRUKOW, G. 1964. *Z. Tierpsychol.* 21:279–301.—
BURNS, E. R., L. E. SCHEVING, and TIEN-HU TSAI. 1972. *Science* 175:71–73.
DAVIS, W. M., and O. L. WEBB. 1963. *Med. Exptl. (Basel)* 9:263–267.
EMLEN, S. T., and W. KEM. 1963. *Science* 142:1682–1683.—
ERTEL, R. J., F. HALBERG, and F. UNGAR. 1964. *J. Pharm. a. Exp. Therapeutics* 146:395–399.
FORWARD, E. F., and D. DAVENPORT. 1970. *Planta (Berl.)* 92:259–266.
GIBSON, R. N. 1970. *Anim. Behav.* 18:539–543.
HALBERG, F., E. JOHNSOHN, W. BROWN, and J. J. BITTNER. 1960. *Proc. Soc. exp. Biol. (N.Y.)* 103:142–144.—
HAUS, E., and F. HALBERG. 1959. *J. appl. Physiol.* 14:878–880.—
HEATH, W. G. 1963. *Science* 142:486–488.—
HOFER, K., u. R. BIEBL. 1965. *Protoplasma* 59:506–521.
JACKLET, J. W. 1971. In *Biochronometry*. M. MENAKER, ed. Pp. 351–362. Washington, D.C.: Nat. Acad. Sci.—
JAHN, T. J., and F. CRESCITELLI. 1940. *Biol. Bull.* 78:42–52.
KAISER, I. H., and F. HALBERG. 1962. *Ann. N. Y. Acad. Sci.* 98:1056–1068.
LINDBERG, R. G., J. J. GAMBINO, and P. HAYDEN. 1971. In *Biochronometry*. M. MENAKER, ed. Pp. 169–185. Washington, D.C.: Nat. Acad. Sci.—
LUTSCH, E. F., and R. W. MORRIS. 1967. *Science* 156:100–102.
MEIER, A. H. 1969. *Gen. a. comp. Endocrinol. Suppl.* 2:55–62.—
MENG, K., u. K. POHLE. 1961. *Z. Krebsforsch.* 64:219–223.—
MENZEL, W. 1952. *Z. Alternsforsch.* 6:104–121.—
MOTTRAM, J. C. 1965. *J. Path. Bact.* 57:265.
NEUSCHELER, W. 1967. *Z. Pflanzenphysiol.* 57:151–172.
PALMER, J. D. 1964. *Amer. Naturalist* 98:431–434.—
PIZZARELLO, D. J., D. ISAAK, and KIAN ENG CHUA. 1964. *Science* 145:286–291.—
POHLE, K., E. MATTHIES, u. K. MENG. 1961. *Z. Krebsforsch.* 64:215–218.—
POLCIK, B., W. NOWOSIELSKI, and J. A. NAEGELE. 1964. *Science* 145:405–406.
REMMERT, H. 1960. *Biol. Zentralbl.* 79:577–584.—
RENSING, L. 1969. *Z. vergl. Physiol.* 62:214–220.—
RIMET, M. 1900. *Ann. Biol. (Paris)* Sér. 3, 64:189–198.
SCHEVING, L. E., D. F. VEDRAL, and J. E. PAULY. 1968. *Anat. Rec.* 160:741–749.—

Schwemmle, B. 1960. *Cold Spring Harbor Symp. quant. Biol.* **25**:239–243.–
Schwemmle, B., u. O. L. Lange. 1959. *Planta (Berl.)* **53**:134–144.–
——. 1959. *Nachr. Akad. Wiss. Göttingen II*, Jg. 29–35.–
Spalding, J. F., and P. McWilliams. 1965. *Health Phys.* **11**:647–651.–
Stanners, C. P., and J. E. Till. 1960. *Bioch. Bioph. Acta* **37**:406–419.

13. Use of the Clock for Day-Length Measurement

> A certain kind of caterpillar is hidden in cracks after the sun has begun to recede from the summer tropic and it . . . becomes surrounded by a hard, horny, annular skin.
>
> ALBERTUS MAGNUS: De Animalibus (13th Century). Quoted from A. C. CROMBIE: Augustine to Galileo. London: Falcon Press 1952.

a. Survey of day-length measurements

General comments and historical facts. Many organisms use day-length measurements to orient themselves to the change of seasons. The control of physiological reactions by measurement of day length is known as "photoperiodic control."

In earlier times, the opinion prevailed that annual changes in temperature, light intensity, precipitation, etc., were responsible for the adjustment by organisms of the course of their development to seasonal changes. This is, indeed, generally true, but only when the organism can place sufficient reliance on such a change as an indicator of the time of year. For example, the temperature changes in some oceans are good indicators of the time of year, and the cycle of development in certain closely investigated marine organisms (especially, algae) have proved to be primarily under the control of temperature (D. MÜLLER). In other cases, however, temperature is not a reliable measure of the course of the seasons; it is then that photoperiodic control is often effective. In photoperiodic responses, organisms measure the real day (or night) length, i.e., approximately the time from dawn to dusk (or vice versa) without considering the naturally occurring fluctuations in the intensity of the light.

The first indications of a photoperiodic control of physiological processes were obtained during studies on flower formation in plants. HENFREY (1852) mentioned day length as a possible factor in the distribution of plants, TOURNOIS (1912) as a factor in flower initiation. KLEBS (1913) expressed his results with *Sempervivum funkii* somewhat more precisely:

"In nature the time of flowering is very probably determined by the fact that day length increases after the spring equinox (March 21st). After the day reaches a certain length, the formation of flowers is initiated. Light probably does not function as a nutrient factor, but more in a catalytic way."

At first the true nature of the photoperiodic reaction was not recognized. People thought that this was something specifically connected with flowering. Only about two decades later did experiments make apparent the existence of a very widely occurring principle of control: a block *may* be incorporated into a variety of physiological processes in plant and animals, and this block is controlled by a time-measuring process (just as a switch or block controlled by a clock *may* be incorporated into different technical devices, machines, illumination and heating systems, etc.).

In spite of the diversity of physiological processes photoperiodically controlled in some plant or animal species, there are conditions or varieties known in which each one of these processes also functions independently of the photoperiodic control. There are, for example, "day-neutral" plants that flower independently of day length. Evidently, the photoperiodic control of developmental processes is especially well developed where it is ecologically important, i.e., where it offers selective value. It became especially important during the conquest of dry land, where temperatures are a much less reliable indicator of seasonal changes than in the ocean. We should not be surprised to discover that photoperiodic control is found most frequently in organisms living between 35 and 40 degrees latitude, as JUNGES has shown for plants. In this area temperatures are the least reliable as a criterion of seasonal changes. By contrast, in tropical areas, which are continuously moist, information about seasonal changes is usually entirely superfluous.

MARSHALL quips with respect to the photoperiodic control of migration and reproductive cycles in birds: "Light is important only in species for which it is important that it should be important."

For example, seasonal changes in testicular size are very often controlled by seasonal changes in day length. However, the equatorial weaver finch *Quelea quelea* shows the seasonal cycles also when kept under an unchanging daily 12-hour photoperiod (LOFTS 1964).

It is also not surprising that photoperiodic control of developmental cycles often degenerates in domesticated animals (ORTAVANT et al.).

The photoperiodic control of reproductive cycles can also be disrupted by certain external factors, including extreme temperatures (see p. 203). Chemical factors can eliminate photoperiodic control of flower formation (HILLMAN 1962; KANDELER).

Attempts to explain photoperiodic time measurement with the help of circadian rhythms originated from earlier discussions on the possible selective value of these rhythms. WEISMANN (1906) postulated such a selective value, but he could not give a satisfactory answer. Much later it was assumed that differences in the periods of inherited and external 24-hour cycles might result in physiological disturbances, thus giving the circadian rhythm a selective advantage (BÜNNING 1932). This is now well known to be true (p. 230). Somewhat later (BÜNNING 1936), the more special hypothesis on time measurement in photoperiodism was suggested.

Botanical examples. Since the investigations by GARNER and ALLARD (1920), it has become a widespread practice to treat all questions of flowering as photoperiodic problems, and even to refer to them this way. This is very misleading. Flowering, under normal or under experimental conditions, will often occur completely independently of day length. One prerequisite for photoperiodic control of flowering is that a correlative interaction exist between the leaves that perceive the day length and the apex that actually changes into the flower. The leaves exert influences on the apex, which promote or inhibit flowering. The length of day or night determines whether this physiologically important condition of the leaves will materialize. The same physiological condition of the leaves that is controlled by day length can also influence completely different processes of development in the plant. Questions about the characteristics of the so-called flowering hormone or about the role of growth substances will not be discussed. *We are interested here only in the time-measuring ability that finds its expression in photoperiodic responses.*

We find this phenomenon in completely different physiological reactions; i.e., day length influences various processes of growth, development, and behavior in plants and animals. Plant species are known in which at least one of the following processes depends on day length: tuber formation, seed germination, vegetative development, succulence, cambium activity, tissue differentiation and induction or termination of dormancy of buds, bulbs, etc.

Long-day plants form flower primordia after being exposed for a few days to light periods of, for example, more than 12 hours. If the light period is shorter than this critical day length of 12 hours, no flower primordia at all will be formed. In a few long-day plants even one long day is sufficient for induction. So-called short-day plants can grow vegetatively from several months to an unlimited time, if the light period is longer than, for example, 12 hours. Yet if the light period is shorter than the critical day length for a few days only (in some species a single short day is sufficient), then flowering rapidly follows.

Use of the Clock for Day-Length Measurement

Zoological examples. SCHÄFER suggested in 1907 that day length might be the controlling factor in the migration of birds as well as in the annual gonad cycles. The experimental approach to animal photoperiodism started later.

MARCOVITCH recognized in 1924 the photoperiodic control of the known annual cycle in the behavior of plant lice. The appearance of sexual animals and the deposition of eggs in *Aphis forbesi*, which is usually restricted to autumn, can be attained in spring simply by shortening the day length. In *Aphis sorbi*, a prolongation of the light period in autumn causes migration, which usually takes place only in spring. ROWAN (1926 and later) reported that several types of animal behavior, occurring only during a certain time of the year, show a dependence on day length. Increasing the daily light period in autumn can induce the enlargement of the gonads and the impulse to migrate northward in *Junco hyemalis, Cervus brachyrhynchos,* and *Serinus canarius.* These phenomena are otherwise restricted to spring.

KOGURE (1933) discovered photoperiodic control of diapause in the commercial silkworm (Fig. 118b). This effect of photoperiod on induction or termination of diapause occurs in many insect species (see LEES *1955, *1960; DE WILDE, 1962; *DANILEVSKII; *DANILESKY *et al.*, *BECK; *ADKISSON). (Diapause, a rest period comparable to dormancy of buds or bulbs in plants, is characterized by complete or nearly complete reduction of many metabolic processes.)

In later zoological works, photoperiodic control of various other phenomena was recognized and examined more closely. Some examples are migratory restlessness, development of sex glands, mode of reproduction, change of fur color (e.g., weasel), wool growth in sheep (HART 1961; HART *et al.*; L. R. MORRIS; BENNETT *et al.*), and molting (see also *WOLFSON; *ASCHOFF (ed.); *FARNER; *LOFTS; *LEES; *BENOIT et ASSENMACHER (eds.); *FARNER and LEWIS).

Critical day length. The decisive factor in photoperiodic reactions normally consists in exceeding a critical day length, regardless of whether the examined process of development is being promoted or inhibited by the increased day length. The length of this critical light period varies; usually it is between 10 and 14 hours. Fig. 117 shows the difference between the two types of relation between day length and physiological response. "Physiological response" may be flower initiation or one of the several other examples of dependence on the day length in plants and animals. Two special examples are given in Fig. 118.

The value of this critical day length must vary with the geographical latitude in order to make it a suitable measure for finding the seasons. This has been confirmed several times. For example, diapause

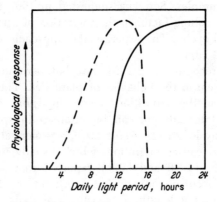

Fig. 117. The two types of the possible relation between day length and physiological response. The diagram shows one reaction requiring a minimum day length of 11 hours and another reaction limited by a maximum day length of 16 hours.

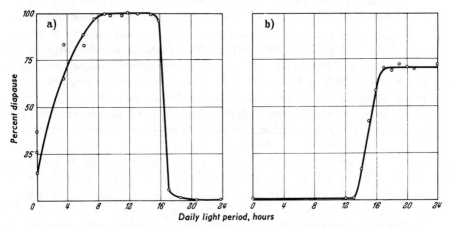

Fig. 118. Photoperiodic induction of diapause in two butterflies. (a) *Acronycta rumicis* (after DANILEVSKII), (b) *Bombyx mori* (after KOGURE). (a) and (b) from LEES 1959

in different geographical populations of the butterfly *Acronycta rumicis* is induced by day lengths of less than about 15 hours in populations from 43°N, but by day lengths of less than 18 hours in populations from 50°N (*DANILEVSKII; for geographical strains of *Pectinophora gossypiella*, see ANKERSMIT and ADKISSON). In a strain of wild rice, the critical day length becomes on the average 3.5 minutes longer when its native habitat lies one degree more northward in the Northern Hemisphere or southward in the Southern Hemisphere (KATAYAMA).

b. Accuracy and reliability of day-length measurements

Accuracy. The curves referred to above, demonstrating the relation between day length and rate of the respective physiological process, indicate the accuracy with which plants and animals are able to measure day length. Often, the beginning of migration in birds can be predicted regardless of the prevailing weather, with an error of only a few days. Obviously, birds are able to distinguish differences of a few minutes in day length and plants are able to measure the day length just as accurately. For example, there are rice varieties responding to differences in day length of only 48 minutes with differences in rates of development of 50 to 60 days (KERLING). This result, including its quantitative relations, has been confirmed repeatedly (see NJOKU; KATAYAMA). Thus, a difference in day length of only one minute results in an acceleration or inhibition of development for more than one day. Examples such as this indicate that in plants a photoperiodically controlled orientation to the seasons is possible, even in areas close to the equator, in spite of the rather small seasonal changes in day length in those areas.

It should not be surprising, therefore, that even animals living close to the equator (e.g., bats from the New Hebrides, according to BAKER) may follow the seasons in their reproductive cycles.

Reliability over wide ranges of temperatures. As in any biological time measurement, the question arises as to whether the organisms can rely upon the time-measuring process in photoperiodic control, or whether a temperature error might be possible. As we have seen, the error caused by temperature is very small or practically absent in physiological processes that depend on the circadian clock, since the period is almost independent of temperature.

Surprisingly enough, time measurement in photoperiodic reactions is also independent of temperature to a large extent. In other words, temperature has only a slight influence on the critical day length. This is shown in botanical and zoological examples in Figs. 119 and 120.

For further examples, see PARIS and JENNER (induction of diapause in *Metriocnemus knabi*); STROSS and HILL (induction of diapause in *Daphnia pulex*); LEES (1963, induction of diapause in the aphid *Megoura viciae*); ANKERSMIT and ADKISSON (induction of diapause in the moth *Pectinophora gossypiella*); *DANILEVSKII (induction of diapause in butterflies). DANILEVSKII also cites examples of a somewhat greater influence of temperature. This influ-

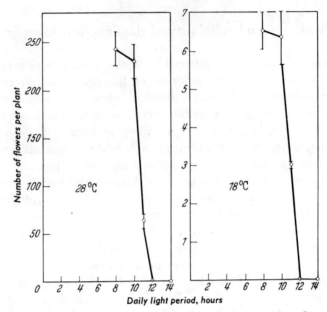

Fig. 119. *Kalanchoe blossfeldiana*. Effect of daily light period on flower formation. The temperature influence on the critical day length between 18 and 28°C is only very slight. Note the very strong influence of temperature on quantity of flowers. Original

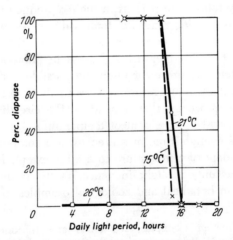

Fig. 120. *Pieris brassicae*. Effect of daily light period on the percentage of diapause at various temperatures. The temperature influence on the critical day length between 15 and 21°C is only very slight. Higher temperature prevents photoperiodic control. After BÜNNING u. JOERRENS

ence may be traced to an influence of temperature on the phase angle difference between the temperature cycle and the circadian cycle.

The total lack of, or only slight influence of, temperature on the critical day length has often been overlooked, because the temperature influence on the controlled process (e.g., flowering, diapause) has been confused with the temperature influence on the critical day length. It is often possible (and perhaps this always holds true) that extremely high or low temperatures entirely exclude the photoperiodic control. Such extreme conditions thus eliminate the block, which is usually released only by the time-measuring process, somewhat like a short circuit in a time switch. Sometimes, the extreme temperature may set an additional block, but as long as the photoperiodic reaction is not excluded by extreme temperatures, it becomes apparent that the critical day length is essentially independent of temperature. It is possible that besides day length, temperature may also have a direct influence on the developmental process in question. Such a direct influence becomes very obvious in Fig. 119, which shows the lack of temperature influence on the critical day length but the strong influence on the quantity of flowers.

> Ecologically, it may be advantageous for the organism to exclude the photoperiodic control during extreme temperatures. Consider, for example, larvae of the cabbage butterfly, which are exposed to very high temperatures during short days. It may be advantageous for the larvae not to develop into diapause pupae in spite of the short days because extremely high temperatures together with relatively short days indicate to them that they are living in subtropical regions, where a day length of 13 to 14 hours does not imply the approach of winter. The exclusion of photoperiodic control and its replacement by temperature control with high temperature has been observed several times (see LEES 1963; PARIS and JENNER). In the same way, it may be advantageous for some organisms that low temperatures always initiate a diapause or a similar physiological state, even in long days. In general, it must again be emphasized that the controlling influence of a certain factor in physiology, as in engineering, does not exclude the controlling influence of certain other factors.

As has been pointed out, time measurement in the photoperiodic reaction is independent of temperature to a large extent. This suggests the assumption that here, too, time is measured by means of the circadian clock.

Reliability over wide ranges of light intensities. Photoperiodic information about seasonal changes has to be reliable. Therefore, the relation between day length and intensity of the physiological reaction must be valid not only for high light intensities, but also when the light intensities have decreased to a small fraction of the "normal."

The threshold of the photoperiodic time-measuring process is usually quite low. This process begins when the sun is still a few degrees below the horizon; it ends in the evening, when the sun has descended a few degrees below the horizon (see PARIS and JENNER). The threshold is, however, usually still above the intensity of moonlight (with certain exceptions), i.e., moonlight generally cannot interfere too much with the photoperiodic orientation to the seasons.

The threshold for the photoperiodic diapause induction of the Colorado potato beetle (*Leptinotarsa decemlineata*) and of other insects is at about 0.1 lux. The reaction depends upon the light intensity only up to 5 lux (DE WILDE and BONGA). This means that as low as 5 lux, or less than 0.01% of full sunlight at noontime, still induces the full reaction. In temperate zones, the intensity of full moonlight does not reach more than 0.25–0.5 lux.

Measurements in plants reveal similar threshold values for the photoperiodic control of flower initiation (see KATAYAMA on the rice plant and for additional references). Thus, plants as well as animals actually measure the time from morning to evening civil twilight (i.e., between light intensities of about 1–3 lux, or between sun positions of about 6–7° below the horizon. As pointed out, the situation is very similar with respect to light intensities for synchronizing the circadian rhythmicity (Chapter 6(b)).

Plants as well as animals may show special anatomical structures allowing sufficient light absorption during twilight periods. The pupae of the oak silkworm *Antheraea pernyi* have a window-like transparent zone directly over the region of the brain, which is the receptor of the photoperiodic stimulus. This transparency is effective especially for blue light (WILLIAMS *et al.*). The bud scales of the beech tree have a transparent basal portion that allows sufficient light to reach the leaf-primordium (WAREING). These primordia are the receptors for the light that terminates the dormancy of the bud. Leaves of plants are usually the site of light perception in the photoperiodic control of flowering. Apparently, the epidermis plays a decisive role in this perception (references: BÜNNING and MOSER 1969). The epidermis does not contain chlorophyll. This fact, in connection with specific optical properties of the epidermis, makes bringing light to the decisive pigment highly efficient. (BÜNNING 1972).

c. The nature of the time-measuring process

Effects of light interrupting the dark period. Analysis of the photoperiodic phenomena has shown that it is not the length of the day as such that is decisive. The effects under consideration (promotion or inhibition of a certain stage of development) can also be obtained by giving short days instead of long days and dividing the nights into two dark periods by a relatively short supplemental light (e.g., 1 hour or much less) (Fig. 121). Other evidence has also led to the conclusion that it is not the length of the day (i.e., the length of the light period) that is decisive, but the length of the night. One possible interpretation of these findings is that a time-measuring process is initiated at the beginning of the light period or at the beginning of the dark period. At a certain number of hours after the beginning of light or darkness this process induces a sensitive stage that responds specifically to light. In fact, it can be demonstrated that this interpretation is correct. There is ample proof that the strongest effect is not obtained by simply dividing the dark period into two *equal* parts with such a light interruption. Rather, the time of highest sensitivity has a disinct relation either to the beginning of the light period or to the beginning of the dark period. This may be asymmetric, depending upon the length of the dark (or light) period (Fig. 122, see also ADKISSON and *ADKISSON).

These effects of light interruptions, i.e., obtaining long-day effects by interrupting the dark period after a short day, have been observed in plants and animals of very different taxonomic groups. Some more examples from vertebrates will be added to the observations made on plants and insects. HART (1951) investigated the initiation of the breeding period in the ferret. Twelve hours of light each day for about 2 months can initiate this period, if 1 hour of light is given from midnight to 1 A.M. It required 18 hours of light daily if the light was offered as a continuous light period. Even 6 hours of light daily are sufficient to bring the ferret into breeding condition when the

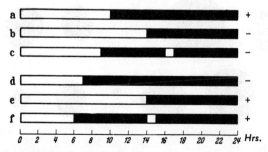

Fig. 121. Interrupting the dark period results in long-day effects in spite of short-day conditions. Flowering in short-day plants (a-c) and in long-day plants (d-f) +: flowering, —: not flowering.

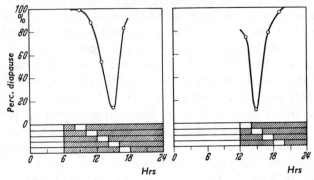

Fig. 122. *Pieris brassicae*. Diapause inhibiting effects of light breaks (2 hours) during dark period. Both with a relatively long and a much shorter main light period. Dark periods indicated by shading. There is a maximum sensitivity about 14–16 hours after the beginning of the main light period. After BÜNNING u. JOERRENS

cycle consists of 4 hours of light, and a 20-hour dark period interrupted by light 17 to 19 hours after beginning of the light period (HAMMOND). Experiments with birds are shown in Fig. 123. A greater coincidence with the response of plants could not be expected.

Similar results are known for birds (Fig. 119). To produce a certain amount of sperm, male juncos (*Junco hyemalis*) had to be ex-

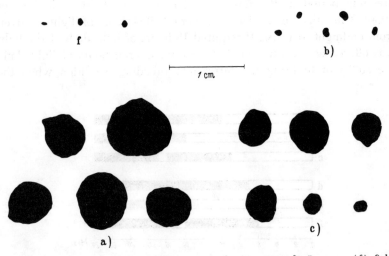

Fig. 123. *Junco hyemalis*. Relative sizes of testes in early January. (f) field controls at start of experiment, (a) after 55–57 days of long-day treatment. (b) short-day control; (c) short-day with interrupted nights. After JENNER and ENGELS

posed either to 14 hours of light each day or to only 8 hours, provided that 2 hours of the 8 were given as an interruption of the dark period 16 to 18 hours after the beginning of the light period. Sparrows (*Zonotrichia leucophrys gambelii*) responded the same way (JENNER and ENGELS). The egg-laying activity of chickens is also a good example. These observations lead to conclusions coinciding with those reported for plants (see also ASCHOFF 1955; FARNER and MEWALDT).

Light interruptions of less than 5 minutes may cause these results. This holds true for both plants (see, e.g., PARKER *et al.*) and animals (see, e.g., BARKER; ANKERSMIT). These authors suggest the application of this effect in the experimentation of insect pests. Thus, the prevention of diapause by a brief supplementary light break offered at the appropriate time would result in exposing the nonresistant larvae or pupae to the severe winter temperatures. Besides determining the effects of light interruptions, several objects have also been tested for changes in sensitivity throughout the entire dark period, by offering light breaks at different times during the dark period. This has revealed quite clearly the increasing and decreasing sensitivity, which might be expected in a cyclical physiological process controlled by the circadian clock.

> For further experiments demonstrating changes of responsiveness to light breaks during the dark period see LEES (1965, on diapause induction in the aphid *Megoura viciae*), °FARNER (1965, on testicular response of the white-crowned sparrow *Zoonotrichia leucophrys gambelii*), °MENAKER (1965, on testicular response in *Passer domesticus*).

"Hourglass" processes or the circadian clock? The facts already mentioned point to the involvement of cycles that have many features in common with the cycles of circadian rhythms. Those facts do not yet show explicitly that the cycles are periods of a circadian rhythmicity. They might be equally well "hourglass" processes. But, as we have seen in other diurnal processes, it can also be demonstrated for the sensitivity to light that a LD cycle does not induce only *one* cycle of these changes in responsiveness. Under certain conditions, it induces several cycles. This demonstration requires basically the same kind of experiment described earlier—the transfer of plants into DD for several days and short light interruptions at various times, demonstrating that the periodic changes in quantity and quality of light sensitivity may continue for 2 to 3 days (Figs. 124, 125, 128, 129). For similar experiments with other plants, see HAMNER and TAKIMOTO, also COULTER and HAMNER; with birds, W. M. HAMNER; with insects, SAUNDERS.

Hourglass behavior. In some cases, there is no indication of a continuation of cyclic changes of responsiveness to light for several days.

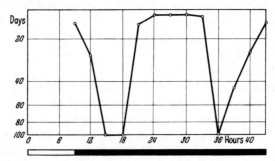

Fig. 124. *Kalanchoe blossfeldiana.* Effect of light breaks (2 hours) at various intervals during long dark period. Times of maximum inhibition about 24 hours apart. Horizontal strip below indicates 9:39 hour LD cycles. Ordinate: days required until flowering. After Bünsow

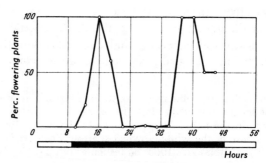

Fig. 125. *Hyoscyamus niger* (long-day plant). Effect of light breaks (2 hours) at various intervals during long dark period. Times of maximum promotion 24 hours apart. Controls (without light breaks) are not flowering. Horizontal strip below indicates 9:39 hour LD cycles. After experiments of Claes u. Lang

This has been stressed especially by Lees (*1966, *1968, *1972), whose extensive experiments with *Megoura viciae* seem to demonstrate clearly that this aphid does not use a circadian oscillation in responsiveness to light. The same conclusion may be drawn from experiments with several other insects (e.g., Fig. 126).

We cannot and should not exclude the possibility that different time-measuring systems are involved in the several species. On the other hand, the conformity of several features in many cases may suggest that the hourglass behavior is caused by a quick damping of the rhythm in longer dark periods. There are several facts favoring this explanation.

Circadian rhythms of leaf movements, running activity, etc., may continue for several weeks in certain cases, but for a few days or only

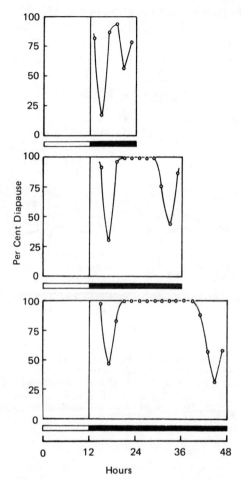

Fig. 126. *Pieris brassicae.* Percentage diapause after LD cycles of 12:36 hours. Light breaks of 30 minutes were offered at different times in the dark period. From *Bünning 1969

about one day in other cases. This depends both on genetic constitution and on special external conditions. For example, the running activity of the beetle *Carabus problematicus* may be fully rephased by a new light program within 4 days, or even within one day, depending on the light intensity (for references, Bünning 1969).

In many cases, authors of earlier studies on plants kept for 2 or 3 days in DD were searching in vain for such results as shown in Figs. 128 and 129. Instead, they found just the same type of reaction as observed in the *Megoura* and several other insects. An example for such an hourglass behavior is given in Fig. 127. The example shows the rapid damping of two circadian oscillations: leaf movements as well as photoperiodic responsiveness. Plants usually react in this way when offered long dark periods following a light period with rather high intensity, or with too much of far-red. Only with improved experimental conditions was it possible to prevent this damping (Fig. 128).

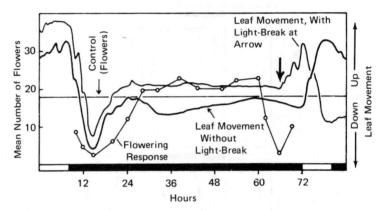

Fig. 127. *Glycine max* (soybean). Synchronous changes in photoperiodic responsiveness to 30-minute light breaks during a long dark period and in leaf movements. The leaf movements also refer to the 8:64 hour LD cycle, but without light break. Natural daylight of the greenhouse was offered before starting this special treatment. Strong damping in circadian leaf movement and in circadian fluctuations of photoperiodic responsiveness ("hourglass" behavior). From °BÜN-NING 1969

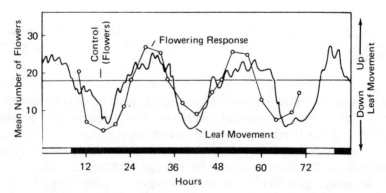

Fig. 128. Similar to Fig. 127, but plants in artificial light (fluorescent tubes, about 10,000 lux) before starting the special treatment. The circadian rhythmicity in responsiveness to light as well as in leaf movements remains clear. From °BÜN-NING 1969

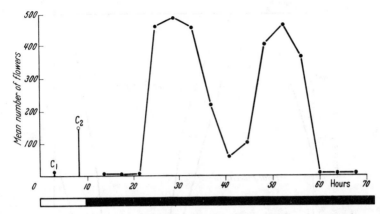

Fig. 129. *Kalanchoe blossfeldiana*. The plants were offered 10:62 hour LD cycles. The dark period was interrupted by light breaks of 1 hour. Time of light breaks on abscissa. Horizontal strip below indicates 10:62 hour LD cycles. Diurnal fluctuation of responsiveness to light. C_1 in continuous light, C_2 in 10:62 hour LD cycles without light breaks. After MELCHERS

Thus the fact that these rhythmic fluctuations in light sensitivity do not continue for several days in all experiments is not a valid argument against the interpretation mentioned above. It is rather exactly what would be expected.

On the other hand, it is also clear from studies with plants that the circadian rhythmicity is not the only process involved in seasonal time measurement in photoperiodism. The additional role of hourglass processes is well known. But the described accuracy and reliability of day-length measurement referred to are mainly caused by the involvement of the circadian clock (for details, see *EVANS; KING and CUMMING).

Correlations between rhythms of responsiveness to light and other circadian rhythms. We know from phenomena of dissociation or of coupling and uncoupling described earlier that the continued expression of one circadian function in an individual plant or animal does not necessarily indicate the running of another clock-controlled function. But in many cases the circadian fluctuations in responsiveness to light are very clearly correlated to other circadian processes. In plants, such a correlation can be established with respect to leaf or petal movements (Figs. 128, 130). But this is not a general rule (SALISBURY and DENNEY). HILLMAN (1970, 1971) used the circadian CO_2 output as an index of circadian timing in photoperiodism of the duckweed *Lemna*.

The correlations can be so parallel that in some cases the leaf's

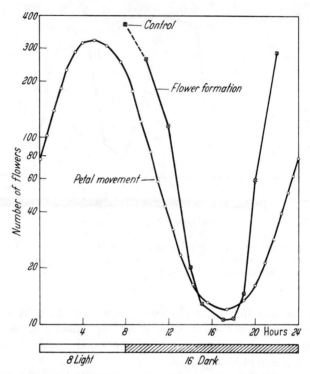

Fig. 130. *Kalanchoe blossfeldiana.* Synchronous diurnal changes in photoperiodic responsiveness to light breaks during the dark period and in the opening degree of flowers. A 5-minute light break has the strongest inhibiting effect on flower dormation when offered 17–18 hours after the beginning of the main light period. This is the same time during which flowers in already flowering plants show maximum closure. These petal movements refer also to a 8:16 hour LD cycle, but without light breaks. Abscissa: hours. Ordinate: number of flowers per plant. No ordinate values are given for the petal movement. Original

position indicates the momentary state of photoperiodic light sensitivity (BÜNSOW 1953, *1960; BÜNNING 1954; BREST et al.; HALABAN; *BÜNNING 1969; Fig. 130).

Other evidence of circadian fluctuations in sensitivity. Experiments with light interruptions are one possible way to demonstrate fluctuations in responsiveness to light. Another experimental approach is that of K. C. HAMNER and coworkers, as applied to plants: the length of the light period is varied over a wide range while the length of the dark period remains constant. In the first instance, any additional light exceeding the optimum light period of 12 hours coincides with that part of the oscillation, in which the process in question is inhibited. If the light period is extended even more, then it coincides with another part of the oscillation, and this further extension has a promoting effect. This means that the quantity and quality of these physiological reactions change periodically during a continuous extension of the light or dark period (*K. C. HAMNER; *HAMNER and TAKIMOTO).

An example of the second instance follows: BLANEY and HAMNER exposed the short-day plant Biloxi soybean to cycles with 8 hours of light each day and dark periods of varying length. Prolonging the dark period up to 16 hours had a favorable effect, further prolongation, however, was inhibitory; but when the dark period was extended still further, to about 40 to 44 hours, flowering was promoted again. This means that during the long dark period the plants endogenously reach a second "photophil" state, and this occurs about 24 hours after the first one (after about 16 hours of darkness). Fig. 131 clearly shows this relation. FINN and HAMNER were able to demonstrate equivalent effects in long-day plants. For similar experiments with plants, see HAMNER and TAKIMOTO; CUMMING et al: with butterflies, BÜNNING u. JOERRENS, with birds, W. M. HAMNER 1964, 1965.

Dual role of light in normal and skeleton photoperiods. The relation between LD cycles and the resulting photoperiodic response is somewhat complex. One of the functions of the LD cycles is to set the phases of the circadian clock. This is known to be due to the specific type of phase shift resulting from offering light to the respective phase, i.e., to the responsiveness of the clock itself which is characterized by the "response curves" (Chapter 6(b)). The other

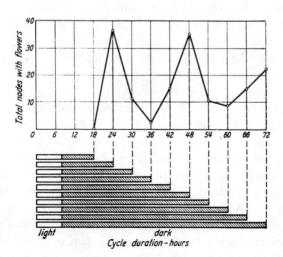

Fig. 131. Flowering response of Biloxi soy-beans to cycles consisting of 10 hours of light and dark periods of 8 to 62 hours. After HAMNER from LANG

role of the LD cycles is due to the fact that the circadian clock on its part causes a rhythm in the responsiveness of the organism to light with respect to developmental processes. Because these two functions are quite often confused, the chapter on rhythmic responsiveness to external factors was placed before this chapter on photoperiodism. The preceding chapter sought to show that the ability of the clock to enforce a rhythm in responsiveness to light, temperature, drugs, etc., is quite different from its ability to become entrained by light signals.

Physiological distinction between the two roles. At least several species of plants and animals can distinguish these two functions of light to a certain degree. They may achieve this by using different organs or tissues for the decisive photoreception, by using different photoreceptors, or different threshold values for the effective light intensities.

A few examples may demonstrate this. The main site of photoreception in plant photoperiodism is the blade of the leaf, but the main site of light absorption leading to phase shifts of rhythm is the leaf joint (BÜNNING u. MOSER 1966, with respect to *Phaseolus*). The threshold values for photoperiodic responses can be lower than those for phase shifts of the rhythm (BREST et al. with respect to the soybean *Glycine max*). In experiments with the house sparrow *Passer domesticus*, MENAKER et al. found that the photoperiodic control of testis growth is mediated entirely by extraretinal photoreceptors in the brain. The eyes do not participate in photoperiodically significant photoreception. But the eyes are involved in synchronizing the circadian rhythm. In this case, the threshold intensities for affecting the phases of the rhythm are much lower (less than 0.1 lux) than those for the photoperiodic response (about 10 lux).

For other birds there is some evidence that both extraretinal photoreceptors and the eyes are involved in the control of photoperiodic responses (GWINNER et al. 1971). Further information on the site of photoreception will be summarized in the section on pigment systems.

There are also examples showing differences in the action spectra for the two light effects. In mediating phase shifts of the circadian rhythm of certain species of plants phytochrome as well as other pigments can be involved. But the cycles of photoperiodic responsiveness, which are controlled by the circadian clock, may be cycles in physiological reactions to light absorbed in phytochrome alone. (For examples and references, see BÜNNING 1969; HALABAN and HILLMAN).

The spatial separation of light absorption for controlling the phases and for photoperiodic responses permits us to assume differences in the action spectra. In birds, the spectral ranges of sensitivity are different for the retina and for the extraretinal regions of the brain (*BENOIT 1964; *BENOIT et ASSENMACHER). In insects, only light from about 350 to about 500 nm has been found to be effective in phase shifts (BRUCE and MINIS). The same limits apply with to light-break effects in insects. However, when colored light (red, or even orange) is offered as a main light period, or added to white light, making this a long day, it has some effect (BÜNNING u. JOERRENS; ADKISSON 1965).

In *Pectinophora gossypiella*, even very low intensities of red light offered as long days are effective in preventing diapause, though they are not effective in synchronizing the known circadian oscillations of this moth (PITTENDRIGH et al. 1970).

Any signal of white light (except for very low intensity) will not only reveal the developmental responsiveness of the organism, but it will also have a certain phase-shifting effect. In simple LD cycles as those to which Figs. 122 and 130 refer, this second effect of light breaks offered in the dark period is not very strong. The main light period dominates in setting the phases. But the situation may become somewhat confusing when longer dark periods are offered, as for example in the experiments in Figs. 125 and 129. Still more extreme is the restriction in experimental programs to strongly reduced "skeleton photoperiods," in which the main light period, normally responsible for the entrainment of rhythms, is reduced to short signals. Under such conditions, the phase-shifting effect of a light break may become relatively strong. Another complicating factor with long dark periods may be a damping of the rhythm. In these cases light breaks have a reinitiating effect. For all these reasons it has been stressed, from the early beginning of experiments with those types of "skeleton photoperiods," that the actual running of the clock under the given conditions should be measured (see, e.g., BÜNNING 1950, 1954, 1955). Diurnal leaf movements may be studied as a hand of the clock for this purpose (see also BÜNSOW). In many cases, however, it is not yet possible "to correlate the behavior of a photoperiodically influenced phenomenon with the concurrent behavior of another system which independently tells us something about the way in which the light regime influences the endogenous clock" (*MENAKER 1965).

> For related experiments and discussions, see HAMNER and TAKIMOTO (1964); HILLMAN (1964); *BECK (1964); *ADKISSON (1964, 1965); FARNER (*1959, *1965); MINIS; *PITTENDRIGH and MINIS (1964); LEES (1965, 1966); *WOLFSON (1965); W. M. HAMNER; CARPENTER and HAMNER; *PITTENDRIGH (1966). Several of the experiments published by these authors stress and clearly demon-

strate the dual effect of light. Also phenomena of damping and reinitiation become clear in several of those publications (e.g., CARPENTER and HAMNER; HAMNER and TAKOMOTO).

d. The pigment systems

General remarks. Another question involves how the circadian clock controls the quantitative and qualitative changes in light sensitivity. The present state of research, however, permits us to comment only on that part of the question that concerns the pigment systems involved. The other part of the question, how they are influenced, remains open.

The sensitivity of very different systems connected with pigments can be modified by the circadian clock. As we have said, the sensitivity of processes connected with eye pigments may fluctuate diurnally. We also pointed out that the photosynthetic capacity can be controlled by the clock. This example in particular shows that the reason for fluctuations in responsiveness to light is not necessarily a fluctuating pigment concentration, for in plants there are only small diurnal changes in chlorophyll concentration, and far from enough to account for the fluctuations in photosynthetic capacity.

The following lines may help to illustrate that quite different pigment systems may indeed be involved. In general, one should probably assume that these fluctuations in reactions to light are not due to changes in pigment concentration. A more decisive consideration must be the complicated influences of the clock on the physiological mechanism, which quantitatively and qualitatively change the ability to cause certain reactions with the help of the absorbed light.

Plants. The so-called red-far-red pigment system (phytochrome) plays a most important role in photoperiodic reactions of plants (at least often and in the most thoroughly investigated cases). (HENDRICKS; *EVANS).

For some time, it has been assumed that diurnal changes in photoperiodic responsiveness are caused only by changes in the proportions of these two forms of phytochrome in the plant. It is now established, however, that the response to one and the same amount of light, either red or far-red, may be quite different even with the same given amount of the two phytochrome forms in the plant (*EVANS).

Phytochrome is clearly a decisive photoreceptor in the photoperiodic response of plants. But the utilization of the products of the photochemical reaction after absorption in phytochrome depends on the phase of the circadian rhythm occurring at the time of that photochemical reaction.

Actually, the statement concerning the decisive role of phytochrome is, in some ways, a simplification. In many cases, light reactions connected with other pigments are also involved. Moreover, it is well known that the light offered to the plant may reveal very different action spectra depending on the time it is offered (extending a short light period, or interrupting the dark period at various times).

Animals. Red light has been found, in many cases, to have a marked effect in different bird species, especially starlings, turkeys, and ducks. Wavelengths between about 440 and 800 nm can be effective. Quite often the effectiveness is restricted to wavelengths above 580 nm (see, e.g., BURGER; *BENOIT; BISSONETTE; RINGOEN; SCOTT and PAYNE). This restriction may be due to the fact that blue light does not penetrate the encephalic region and receptors.

The effective spectral ranges do not coincide with the spectral sensitivity curve of the eye (see Chapter 6(b)). This could be expected in consequence of the role of extraretinal light absorption (p. 214; *BENOIT; *HOLLWICH; *FARNER and LEWIS; *BENOIT et ASSENMACHER; MENAKER et al.; MENAKER and KEATS). Action spectra have not been determined. Thus, the photoreceptor pigments remain still unknown.

There are now many experimental demonstrations of light effects in mammals that are not mediated via the eyes (*HOLLWICH; *BENOIT; see also Chapter 6(b)). Many of these light effects are concerned with the growth of testicles and thus are probably related to the photoperiodic control of developmental processes. The extent to which mammals utilize the different types of photoreceptors in photoperiodism has not yet been studied in detail. It is quite possible that there are species differences in this respect.

The photoperiodic induction of the diapause in arthropods usually seems to be determined by shortwave light (see LEES; DE WILDE; GEYSPITS; BÜNNING u. JOERRENS; H. J. MÜLLER). But insects do not always take up the photoperiodic stimulus by their eyes. LEES suspects that the light may reach the central nervous system through the integument. On the basis of his experiments with the aphid *Megoura viciae*, he came to the conclusion that the photoperiodic receptors are in the protocerebrum. Neurosecretory cells, possibly in the pars intercerebralis, appear to be implicated both as receptors and as humoral effectors. Several other research workers brought additional evidence for photoperiodic control of diapause in insects which is not mediated via the visual organs but rather by light absorption in the brain. (WILLIAMS and ADKISSON; WILLIAMS et al.; ADKISSON; *DANILEVSKII, GRASSÉ). Of special interest are experiments by WILLIAMS and ADKIS-

SON. They excised the brain from the head of the pupa in *Antheraea pernyi* and reimplanted it in the tip of the abdomen. This resulted in a shifting of the photoperiodic responsiveness from the head region to that region of the abdomen. CLARET reported on similar experiments with *Pieris brassicae*.

As mentioned, the situation becomes more complicated because of differences in the action spectra (in certain species) for phase shifts of the rhythm, for effects of light breaks and for light offered during the main light period. Thus, the situation is perhaps not less complicated for animals than it has long known to be for plants. As for plants, we mentioned the dependency of the action spectra on the time within the dark period at which the light is offered. The same applies to the aphid *Megoura viciae* (LEES 1971). Within the daily dark period, there is a stage with maximal sensitivity to 450–470 nm, another stage with a considerable extension of sensitivity into the longer wavelengths.

General conclusion. Very different pigments can participate in light reception of photoperiodic processes. Most of the pigments involved are still unknown. Nothing is known about the immediate pathways following the absorption of light: How is the clock able to control the responsiveness to light?

e. Distinguishing between increasing and decreasing day length

General remarks. Except for the shortest and the longest day, each day length occurs twice a year. How can organisms decide whether a particular day length is in the season of increasing or decreasing day lengths? To do this, the organism must make use of additional information. In other words, the photoperiodically controlled process must be blocked not just at one place, but at two.

> One kind of additional block can be released by low temperatures. This possibility is known for plants as well as for animals. Usually, temperatures of about $+5°C$ are optimum for releasing this block. For example, introducing an additional block in a plant that flowers under long-day conditions may result in a plant that does not flower in the year in which it germinated, but only after the following winter. This type of plant is quite common (e.g., winter wheat). The phenomena of these temperature effects are referred to in botany as vernalization. But there is no reason to consider these botanical phenomena as physiologically different from corresponding low temperature effects in insects or other animals.

Combining two day-length measurements. One possibility is of particular interest: In some cases two blocks have to be released photoperiodically, one by long-day conditions and the other one by short-day conditions. This response type is widespread in the plant kingdom (see, e.g., WELLENSIEK; RESENDE; THOMAS). Usually, the release of one block is required before the second one can be released (only within a certain period after the first one). Sometimes, short-day conditions have to be offered before long days, and in other cases long-day conditions must precede short days. These relations have been studied most thoroughly in plants, yet they are evidently also found in animals. We may refer to the investigations of *WOLFSON in birds (*Junco hyemalis and Zonotrichia albicollis*). Obviously, the day length has a decisive control over the seasonal fluctuations in behavior (swelling of the reproductive glands, migration). Two physiological phases are important. One occurs in late summer and autumn ("preparatory phase"); it is controlled by short days and does not occur under long days. The second phase follows, lasting until beginning of migration in spring, when it is promoted by long days (see also FARNER and MEWALDT). The best egg production of turkey hens was obtained from hens lighted 6 hours a day for 3 weeks, then changed to a 14-hour day (OGASAWARA *et al.*).

In the bollworm *Heliothis zea*, the several developmental stages have different photoperiodic requirements (WELLSO and ADKISSON; ROACH and ADKISSON).

> For further references on the combination of short-day and long-day reactions in insects see KREHAN.
> A similar change from short-day to long-day treatment also results in egg production at an earlier age than in hens offered only a long-day regime (LEIGHTON and SHOFFNER). There is also some evidence for superior development in the domestic fowl when long-day treatment follows a short-day treatment (SIEGEL *et al.*; see, however, also SMITH and NOLES). There may be a similar situation in the photoperiodic induction of diapause in the grasshopper *Nomadacris septemfasciata* (NORRIS), where there are different responses to day length at early and later stages of development.

The situation is still more complex. In birds, for example, a refractory period of several months may occur after the activity of the reproductive glands has been induced photoperiodically. During this period, a new induction of this type is not possible but in some cases even this refractory period can be reduced, apparently by the influence of *other* day lengths. It even appears possible that in some species the refractory period proceeds autonomously (see *WOLFSON; *MARSHALL; MILLER).

A situation as complex as that in birds has been found in lizards: long days promote the growth of testicles and spermatogenesis, but the reaction becomes more intense as the long-day treatment approaches the usual breeding time (Fox and DESSAUER; FISCHER).

These facts may indicate a close relationship between what is called "preparatory phase," "refractory phase," etc. and circannual rhythmicity (see p. 221): those phases might be phases of an endogenous and approximate 12-month cycle.

Responses to the direction of day-length changes. There are several observations indicating that the change in day length in itself is a stimulating factor. According to MORRIS and FOX, the increasing day length is the decisive factor for the rate of sexual maturation in chickens. Here, the actual length of the day does not seem to be decisive, but rather the rate of change of day length. A similar assumption is made by NAGARCENKAR regarding photoperiodic control of wool production in ewes. Any change in the day length stimulates the pituitary and induces increased wool production. But even in these examples it is still possible that the changes in day length are measured by combining two photoperiodic reactions.

One example indicates very clearly the photoperiodic response of an insect to the direction of change in day length (both increasing and decreasing). TAUBER and TAUBER observed in *Chrysopa carnea* (an insect) responses as indicated in Table 8.

Restriction to a narrow range in day length. In organisms in which two blocks have to be released, one by long days and another by short days, the process under control can also be released by constant day length under certain conditions. This is possible when there is a certain overlap between the upper critical day length of the short-day component and the lower critical day length of the long-day component. An example of this is the flowering in *Cestrum nocturnum* (Fig. 132).

Table 8. *Chrysopa carnea*, % diapause in different LD cycles, After TAUBER and TAUBER

LD cycle	Diapause
hr	%
8:16	100
10:14	100
12:12	100
14:10	0
16:8	0
18:6	0
18:6, then 14:10	29
8:16, then 12:12	0

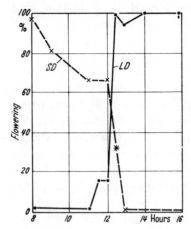

Fig. 132. *Cestrum nocturnum.* Critical day-length for short-day reaction (SD) and for long-day reaction (LD). Both reactions are necessary for flowering. After SACHS

This plant flowers not only after a succession of short days and long days, but also after an exposure to a continuous constant day length. But the possible range of day length between 11½ and 12 hours is effective. Similar examples with a restriction to a rather narrow range of day length have been found over and over again, both in plants and animals.

Sugar cane (*Saccharum spp*) is another example. A certain clone flowers readily with nights of 11 hours 32 minutes and not at 12 hours 26 minutes or above, nor at 10 hours 58 minutes or below (see CLEMENTS, also for further references).

The explanation by a cooperation of a long-day and a short-day reaction holds apparently also in the case of the bollworm, *Heliothis zea* (see p. 219; ADKISSON and ROACH).

Such restrictions as these are prevented if the two different day-length requirements are connected with specific developmental phases, for example with endogenous 12-month cycles ("circannual cycles").

Combination with circannual cycles. It is well known that plants and animals may also have inherent circannual cycles. This phenomenon of endogenous annual periodicity was discovered by botanists and gardeners more than half a century ago in plants taken from Europe to the tropics (*BÜNNING 1956). In such plants, the periodicity of leaf unfolding, flowering, and rest periods of buds persisted, but there was no longer any synchrony. The periods were no longer exactly 12 months and the various developmental phases were totally unsynchronized among different individuals of the same species and even among the different branches of a single tree.

These circannual cycles may, indeed, continue in the laboratory even under constant conditions of day length. The only difference is that such cycles are less accurate than circadian cycles, and there is a tendency toward desynchronization which becomes evident very rapidly (BLAKE 1959; *FRAZER 1959; LOFTS 1964; WOLFSON and WINCHESTER 1959; MERKEL 1963; see also MARSHALL and SERVENTY 1959; *KAYSER; *PENGELLY and ADMUNDSON; *MENAKER [ed.]; GWINNER 1971). Temperature cycles as well as the annual cycles of day length can synchronize the circannual cycles. On the other hand, these circannual cycles can themselves control the photoperiodic responsiveness.

Experiments have confirmed that such external factors as day length or temperature cannot in all cases evoke a specific developmental process at any time. Several plant species show annual fluctuations in the degree of response to the influence of day length. This applies, for instance, to flower induction (CHAUDHRI). An analogous annual fluctuation in responsiveness to the day-length signals has also been found in the growth of gonads in lizards and birds. A specific day length may stimulate the growth of gonads quite easily in certain months, but only very little, or not at all, in other months (FOX and DESSAUER 1958; *FRAZER 1959; SHANK 1959; DOLNIK 1964).

Even qualitative circannual differences in response to day length may occur. Long days may advance gonadal development during certain parts of the circannual cycle, but delay when offered during other phases of this cycle (GWINNER 1971). It is nearly impossible to distinguish between this type of behavior and that previously mentioned, describing the alternation of a period requiring short days with another requiring long days.

Perhaps we are dealing with circannual cycles (or parts of them) in all cases of developmental cycles which include "refractory periods" or "preparatory phases" (p. 219) lasting for a couple of months, during which the organisms do not respond to the photoperiodic stimulus. But this line of reasoning is still in the stage of hypothesis (*FARNER and LEWIS).

f. In retrospect

The facts clearly demonstrate how photoperiodic reactions can be interpreted: as phenomena, which arise only if one or several blocks are incorporated into the complicated mechanism of some develop-

Use of the Clock for Day-Length Measurement

mental process. All these processes, which may be controlled by photoperiodism in some instances (flowering, induction or termination of rest periods, migrations, activity of reproductive glands, etc.), are also possible, in principle, without such a control. In some species, the control may be completely missing, or the control can be eliminated under certain conditions in a single species. The genetic constitutions as well as environmental factors can also determine whether a block will be released by long days, short days, or a combination of both. Considering all this, the great variety of possibilities realized by nature is no longer so amazing as it seemed to be during the early years of research in photoperiodism.

It is no longer surprising to us that certain developmental processes may or may not be coupled to the clock. We have mentioned corresponding phenomena for a wide variety of other physiological processes. Consider animal orientation or mitosis as examples of processes that may or may not be coupled to the clock.

References to Chapter 13
(*Use of the Clock for Day-Length Measurement*)

a. Reviews

*ADKISSON, P. L. 1964. Action of the photoperiod in controlling insect diapause. *Amer. Naturalist* 98:357–374;–

———. 1960. Light-dark reactions involved in insect diapause. In *ASCHOFF, J., ed., pp. 344–350.–

*ASCHOFF, J., ed. 1965. *Circadian Clocks*. Amsterdam: North Holland Publishing Company.

*BECK, S. D. 1963. *Animal Photoperiodism*. New York: Holt, Rinehart and Winston;–

———. 1964. Time-measurement in insect photoperiodism. *Amer. Naturalist* 98:329–346;–

———. 1968. *Insect Photoperiodism*. New York–London: Academic Press.–

*BENOIT, J. 1964. The structural components of the hypothalamo-hypophyseal pathway, with particular reference to photostimulation of the gonads in birds. *Ann. N. Y. Acad. Sci.* 117:204–216.–

*BENOIT, J., et I. ASSENMACHER, eds. 1970. *La photorégulation de la reproduction chez les oiseaux et les mammifères*. Centr. Nat. Rech. Scient. Paris.–

*BORTHWICK, H. A. 1964. Phytochrome action and its time displays. *Amer. Naturalist* 98:347–355.–

*BÜNNING, E. 1956. Endogene Aktivitätsrhythmen. *Encyclopedia of Plant Physiol.* 2:878–907;–

———. 1969. Common features of photoperiodism in plants and animals. *Photochem. a. Photobiol.* 9:219–228.–

°Bünsow, R. C. 1960. The circadian rhythm of photoperiodic responsiveness in *Kalanchoe*. In °Chovnick (ed.), pp. 257–260.

°Chovnick, A., ed. 1960. *Biological Clocks.* Cold Spring Harbor Symp. quant. Biol. 25.–

°Cumming, B. G. 1972. The role of circadian rhythmicity in photoperiodic induction in plants. In *Circadian Rhythmicity.* J. F. Bierhuizen et al. Proc. Intern. Symp. Circ. Rhythmicity Wageningen, pp. 33–85.

°Danilevskii, A. S. 1965. (Identical with Danilesky.) Photoperiodism and Seasonal Development of Insects. Edinburgh: Oliver a. Boyd.–

°Danilesky, A. S., N. I. Goryshin, and V. P. Tyshchenko. 1970. Biological rhythms in terrestrial arthropods. *Ann. Rev. Entomol.* 15:201–244.

°Evans, L. T., ed. 1969. *The Induction of Flowering.* North Melbourne: Macmillan of Australia.

°Farner, D. S. 1959. Photoperiodic control of annual gonadal cycles in birds. In °Withrow (ed.), pp. 717–750;–

———. 1965. Circadian systems in the photoperiodic responses of vertebrates. In °Aschoff (ed.), pp. 357–369.–

°Farner, D. S., and R. A. Lewis. 1971. Photoperiodism and reproductive cycles in birds. In *Photophysiology.* A. C. Giese, ed. Vol. VI, pp. 325–370.–

°Frazer, J. F. D. 1959. *The Sexual Cycles of Vertebrates.* London: Hutchinson Univ. Libr.–

°Hamner, K. C. 1960. Photoperiodism and circadian rhythms. In °Chovnick (ed.), pp. 269–277.–

°Hollwich, F. 1964. The influence of light via the eyes on animals and man. *Ann. N. Y. Acad. Sci.* 117:105–131.

°Kayser, Ch. 1970. Photopériode, reproduction et hibernation des mammifères. In °Benoit et Assenmacher, pp. 409–433.

°Lang, A. 1965. Physiology of flower initiation. *Encyclopedia of Plant Physiol.* 15, 1:1380–1536.–

°Lees, A. D. 1955. *The Physiology of Diapause in Arthropods.* Cambridge: University Press;–

———. 1960. Some aspects of animal photoperiodism. In °Chovnick (ed.), pp. 261–268;–

———. 1968. Photoperiodism in insects. In *Photophysiology.* A. C. Giese, ed. Vol. IV, pp. 47–137. New York–London: Academic Press;–

———. 1972. The role of circadian rhythmicity in photoperiodic induction in animals. In *Circadian Rhythmicity.* J. F. Bierhuizen et al. Proc. Intern. Symp. Circ. Rhythmicity Wageningen, pp. 87–110.–

°Lofts, B. 1970. *Animal Photoperiodism.* London: Edward Arnold Publ. Ltd.

°Marshall, A. J. 1960. Annual periodicity in the migration and reproduction of birds. In °Chovnick (ed.), pp. 499–505.–

°Menaker, M. 1965. Circadian rhythms and photoperiodism in *Passer domesticus*. In °Aschoff (ed.), pp. 385–405.–

———. 1971, ed. *Biochronometry*. Washington, D.C.: Nat. Acad. Sci.—
———. 1972. Nonvisual light reception. *Scientif. Amer.* **226**(3):22–29.
°PENGELLEY, E. T., and S. J. ADMUNDSON. 1971. Annual biological clocks. *Scientif. Amer.* **224**(4):72–79.—
°PITTENDRIGH, C. S. 1966. The circadian oscillation in *Drosophila pseudoobscura* pupae: a model for the photoperiodic clock. *Z. Pflanzenphysiol.* **54**:275–307.—
°PITTENDRIGH, C. A., and D. H. MINIS. 1964. The entrainment of circadian oscillations by light and their role as photoperiodic clocks. *Amer. Naturalist* **98**:261–294.
°WENT, F. W. 1960. Photo- and thermoperiodic effects in plant growth. In °CHOVNICK (ed.), pp. 221–230.—
°WITHROW, R. B., ed. 1959. *Photoperiodism and Related Phenomena in Plants and Animals*. Washington: Am. Ass. Adv. Sc.—
°WOLFSON, A. 1959. The role of light and darkness in the regulation of spring migration and reproductive cycles in birds. In °WITHROW (ed.), pp. 679–716;—
———. 1960. Regulation of annual periodicity in the migration and reproduction of birds. In °CHOVNICK (ed.), pp. 507–514;—
———. 1964. Animal photoperiodism. In *Photophysiology*. A. C. GIESE, ed. Vol. II, pp. 1–49. New York–London: Academic Press;—
———. 1965. Circadian rhythm and the photoperiodic regulation of the annual reproductive cycle in birds. In °ASCHOFF (ed.), pp. 370–378.

b. Other references

ADKISSON, P. L. 1963. *Progress Rep. 2270*. Texas A & M;—
———. 1965. Proc. Conf. on Electromagn. Radiation in Agriculture. New York, pp. 30–33.—
ADKISSON, P. L., and S. H. ROACH. 1971. In °MENAKER (ed.), pp. 272–280.—
ANKERSMIT, G. W. 1968. *Entom. exp. a. appl.* **11**:231–240.—
ANKERSMIT, G. W., and P. L. ADKISSON. 1967. *J. Insect Physiol.* **13**:553–564.—
ASCHOFF, J. 1955. *Studium Gen.* **8**:742–776.
BAKER, J. R., and I. Z. BAKER. 1936. *J. Linn. Soc. (Zool.)* **39**:123–141.—
BARKER, R. J. 1963. *Experientia* **19**:185.—
BENNETT, J. W., J. C. D. HUTCHISON, and M. WODZICKA-TOMASZEWSKA. 1962. *Nature* **194**:651–652.—
BISSONETTE, TH. H. 1943. *Trans. N.Y. Acad. Si. Ser. II* **5**:43–51.—
BLAKE, G. M. 1959. *Nature* **183**:126–127.—
BLANEY, L. T., and K. C. HAMNER. 1957. *Bot. Gaz.* **119**:10–24.—
BREST, D. E., T. HOSHIZAKI, and K. C. HAMNER. 1971. *Plant Physiol.* **47**: 676–681.—
BRUCE, V. G., and D. H. MINIS. 1969. *Science* **163**:583–585.—
BÜNNING, E. 1932. *Naturwiss.* **20**:340–345;—
———. 1932. *Jahrb. wiss. Bot.* **77**:283–320;—
———. 1936. *Ber. dtsch. Bot. Ges.* **54**:590–607;—

———. 1950. *Planta* (Berl.) **38**:521–540;—
———. 1954. *Physiol. Plant.* (Copenh.) **7**:538–547;—
———. 1959. *Ber. dtsch. Bot. Ges.* **67**:421–431;—
———. 1972. In *Circadian Rhythmicity.* BIERHUIZEN et al. Proc. Intern. Symp. Circ. Rhythmicity Wageningen, pp. 9–29.—
BÜNNING, E., u. G. JOERRENS. 1960. *Z. Naturforsch.* **15b**:205–213.—
BÜNNING, E., u. I. MOSER. 1966. *Planta* (Berl.) **69**:101–110;—
———. 1969. *Proc. Nat. Acad. Sci. USA.* **62**:1018–1022.—
BÜNSOW, R. 1953. *Z. Bot.* **41**:257–276.—
BURGER, J. W. 1949. *Wilson Bull.* **61**:211–230.
CARPENTER, B. H., and K. C. HAMNER. 1963. *Plant Physiol.* **38**:698–703;—
———. 1964. *Plant Physiol.* **39**:884–889.—
CHAUDHRI, I. I. 1956. *Beitr. Biol. Pfl.* **32**:451–456.—
CLAES, H., u. A. LANG. 1947. *Z. Naturforsch.* **2b**:56–63.—
CLARET, J. 1966. *Ann. d'Endocrinologie* **27**:311–320.—
CLEMENTS, H. F. 1968. *Plant Physiol.* **43**:57–60.—
COULTER, M. W., and K. C. HAMNER. 1964. *Plant Physiol.* **39**:848–856.—
CUMMING, G., S. B. HENDRICKS, and H. A. BORTHWICK. 1965. *Canad. J. Bot.* **43**:825–853.
DEWILDE, J. 1962. *Ann. Rev. Entomol.* **7**:1–26;—
———. 1965. *Arch. Anat. micr. Morph. exp.* **54**:547–564.—
DEWILDE, J., and H. BONGA. 1958. *Entom. exp. a. appl.* **1**:301–307;—
———. 1958. *Proc. 10th Intern. Congr. Entomol.* **2**:213–218.—
DOLNIK, V. R. 1964. *Zool. Z.* **43**:720–734 (Russian, English summary).
FARNER, D. S., and L. R. MEWALDT. 1953. *Experientia* (Basel) **9**:221–223.—
FINN, J. C., and K. C. HAMNER. 1960. *Plant Physiol.* **35**:982–985.—
FISCHER, K. 1968. *Z. vergl. Physiol.* **60**:244–268;—
———. 1969. *Z. vergl. Physiol.* **61**:394–419.—
FOX, W., and H. C. DESSAUER. 1958. *Biol. Bull.* **115**:421–439.
GARNER, W. W., and H. A. ALLARD. 1920. *J. agr. Res.* **18**:553–606.—
GEYSPITZ, K. F. 1957. *Zoologizeskij J.* **26**:548–559.—
GRASSÉ, M. P. 1966. *C. v. Acad. Sc. Paris* **262**:1464–1465.—
GWINNER, E. 1971. In *MENAKER (ed.), pp. 405–427.—
GWINNER, E. G., F. W. TUREK, and S. D. SMITH. 1971. *Z. vergl. Physiol.* **75**:323–331.
HALABAN, R. 1968. *Plant Physiol.* **43**:1894–1898.—
HALABAN, R., and W. S. HILLMAN. 1970. *Plant Physiol.* **46**:757–758.—
HAMMOND, J. 1952. *J. agr. Sci.* **42**:293–303;—
———. 1953. *Science* **117**:389–390.—
HAMNER, K. C., and A.TAKIMOTO. 1964. *Amer. Naturalist* **98**:295–322.—
HAMNER, W. M. 1965. In *ASCHOFF (ed.), pp. 379–384;—
———. 1964. *Nature* **203**:1400–1401.—
HART, D. S. 1951. *J. exp. Biol.* **28**:1–12;—
———. 1961. *J. Agric. Sci.* **56**:235.—
HART, D. S., J. W. BENNETT, J. C. D. HUTCHINSON, and M. WODZICKA-TOMASZEWSKA. 1963. *Nature* **198**:310–311.—

HENDRICKS, S. B. 1960. In °CHOVNICK (ed.), pp. 245–248.–
HENFREY, A. 1852. *The Vegetation of Europa.*–
HILLMAN, W. S. 1962. *Am. J. Bot.* **4**:892–897;–
———. 1964. *Amer. Naturalist* **98**:323–328;–
———. 1970. *Plant Physiol.* **45**:273–27;–
———. 1971. In °MENAKER (ed.), pp. 251–271.
JENNER, CH. E., and W. L. ENGELS. 1952. *Biol. Bull.* **103**:345–355.–
JUNGES, W. 1957. *Planta (Berl.)* **49**:11–32.
KANDELER, R. 1963. *Ber. dtsch. Bot. Ges.* **75**:431–442.–
KATAYAMA, T. C. 1964. *Jap. J. Bot.* **18**:349383.–
KERLING, L. C. P. 1950. *Proc. kon. ned. Akad. Wet.* **53**:3–16.–
KING, R. W., and B. G. CUMMING. 1972. *Planta (Verl.)*–
KLEBS, G. 1913. *S.-B. Heidelb. Akad. Wiss. Abt. B* **3**:47;–
———. 1913. *Handwörterb. Naturwiss.* **4**:276–296.–
KOGURE, M. 1933. *J. Dept. Agric. Kyushu Univ.* **4**:1–93.–
KREHAN, I. 1970. *Oecologia* **6**:58–105.
LANG, A., u. G. MELCHERS. 1943. *Planta (Berl.)* **33**:653–702.–
LEES, A. D. 1953. *Ann. appl. Biol.* **40**:449–468;–
———. 1964. *J. Insect Physiol.* **9**:153–164;–
———. 1964. *J. exp. Biol.* **41**:119–123;–
———. 1965. In °ASCHOFF (ed.), pp. 351–356;–
———. 1966. *Nature* **210**:986–989;–
———. 1971. In °MENAKER (ed.), pp. 372–379.–
LEIGHTON, A. T., and R. N. SHOFFNER. 1961. *Poultry Sci.* **40**:861–884.–
LOFTS, B. 1964. *Nature* **201**:523.
MARCOVITCH, S. 1924. *J. agr. Res.* **27**:513–522.–
MARSHALL, A. J. 1960. *Symp. Zool. Soc. London* **2**:53–67.–
MARSHALL, A. J., and D. L. SERVENTY. 1959. *Nature* **184**:1704–1705.–
MELCHERS, G. 1956. *Z. Naturforsch.* **11b**:544–548.–
MENAKER, M., and H. KEATS. 1968. *Proc. Nat. Acad. Sci. USA* **60**:146–151.–
MENAKER, M., R. ROBERTS, J. ELLIOT, and H. UNDERWOOD. 1970. *Proc. Nat. Acad. Sc. USA* **67**:320–325.–
MERKEL, F. W. 1963. Proc. XIIIth Internat. Ornithol. Congr. 950–959.–
MILLER, A. H. 1959. *Science* **129**:1286.–
MINIS, D. H. 1965. In °ASCHOFF (ed.), pp. 333–343.–
MORRIS, L. R. 1961. *Nature* **190**:102.–
MORRIS, T. R., and S. FOX. 1958. *Nature* **181**:1453–1454.–
MÜLLER, D. 1962. *Bot. Marina* **4**:140–155.–
MÜLLER, H. J. 1964. *Zool. Jahrb. Physiol.* **70**:411–426.
NAGARCENAKAR, R. 1964. *Photochem. a. Photobiol.* **3**:157–162.–
NJOKU, E. 1959. *Nature* **183**:1598–1599.–
NORRIS, M. J. 1959. *Entomol. exp. appl. (Amst.)* **2**:154–168;–
———. 1965. *J. Insect Physiol.* **11**:1105–1119.
OGASAWARA, F. X., W. O. WILSON, and V. S. ASMUNDSON. 1962. *Poultry Sci.* **41**:1858–1863.–

ORTAVANT, R., P. MAULEON, and C. THIBAULT. 1964. *Ann. N. Y. Acad. Sci.* **117**:157–193.
PARIS, O. H., and CH. E. JENNER. 1959. In °WITHROW (ed.), pp. 601–624.–
PARKER, M. W., S. B. HENDRICKS, H. A. BORTHWICK, and N. J. SCULLY. 1946. *Bot. Gaz.* **108**:1–26.–
PITTENDRIGH, C. S., J. EICHHORN, D. H. MINIS, and V. G. BRUCE. 1970. *Proc. Nat. Acad. Sci. USA* **66**:758–764.
RESENDE, F. 1952. *Portugal. Acta Biol. A* **3**:318–322.–
RINGOEN, A. R. 1942. *Physiol. Zool.* **71**:99–109.–
ROACH, S. H., and P. L. ADKISSON. 1970. *Journ. Insect Physiol.* **16**:1591–1597.–
ROWAN, W. 1926. *Proc. Boston Soc. Nat. Hist.* **38**:147–189.
SACHS, R. M. 1956. *Plant Physiol.* **31**:185–192;–
———. 1959. In °WITHROW (ed.), pp. 315–320.–
SALISBURY, F. B., and A. DENNEY. 1971. In °MENAKER (ed.), pp. 292–311.–
SAUNDERS, D. S. 1970. *Science* **168**:601–603.–
SCHÄFER, E. A. 1967. *Nature* **77**:159–163.–
SCOTT, H. M., and L. F. PAYNE. 1937. *Poultry Sci.* **16**:90–96.–
SHANK, M. C. 1959. *The Auk* **76**:44–54.–
SIEGEL, H. S., W. L. BEANE, and C. E. HOWES. 1963. *Poultry Sci.* **42**:1359–1368.–
SMITH, R. E., and P. K. NOLES. 1963. *Poultry Sci.* **42**:973–982.–
STROSS, R. G., and J. C. HILL. 1965. *Science* **150**:1462–1464.
TAUBER, M. J., and C. A. TAUBER. 1968. *Science* **167**:170.–
THOMAS, R. G. 1961. *Nature* **190**:1130–1131.–
TOURNOIS, J. 1912. *C. R. Ac. Sci. (Paris)* **155**:297–300.
WAREING, P. F. 1953. *Physiol. Plant.* **6**:692–706.–
WEISMANN, A. 1906. *Arch. Rassenbiol.* **3**:1.–
WELLENSIEK, S. J. 1960. *Meded. Landbouwhogeschool Wageningen* **60**:1–18–.
WELLSO, S. G., and P. L. ADKISSON. 1966. *J. Insect Physiol.* **12**:1455–1465.–
WILLIAMS, C. M., and P. L. ADKISSON. 1964. *Biol. Bull.* **127**:511–525.–
WILLIAMS, C. M., P. L. ADKISSON, and C. WALCOTT. 1965. *Biol. Bull.* **128**:497–507.–
WOLFSON, A., and D. P. WINCHESTER. 1959. *Nature* **184**:1658–1659.

14. Pathological Phenomena

Es glaubt nehmlich mancher, es sey völlig einerley, wenn man diese 7 Stunden schliefe, ob des Tages oder des Nachts. Man überläßt sich also Abends so lange wie möglich seiner Lust zum Studiren oder zum Vergnügen, und flaubt es völlig beyzubringen, wenn man die Stunden in den Vormittag hinein schläft, die man der Mitternacht nahm. Aber ich muß jeden dem seine Gesundheit lieb ist, bitten, sich für diesem verführerischen Irrthum zu hüten.

C. W. HUFELAND, Die Kunst das menschliche Leben zu verlägern. 2. Aufl., Jena 1798.

(Many people believe it should not make any difference whether they have their 7 hours of sleep during the day or at night. They pursue their studies or amusements into the night, believing that this may be compensated by extending the sleep into the forenoon. Those who want to maintain their health, should guard themselves against this seductive error.)

a. Disturbances under the influence of nondiurnal rhythms of the environment

General remarks. As we have seen, in some plants and animals the circadian clock controls the diurnal changes in responsiveness to light and temperature. The phenomena of photo- and thermoperiodism show that a normal development of many plants and animals is only possible if the LD cycles or cycles of high and low temperature approximate the 24-hour rhythm. Exposing these organisms to external cycles that deviate considerably in length from the 24-hour cycle inevitably results in disturbances, the origin of which varies from case to case. By offering abnormal cycles, light or darkness or high temperature may coincide with a physiological phase adjusted to the opposite of the prevailing conditions. We also know (see Chapter 6(d)) that the physiological periodicity can only follow the external periodicity within certain limits, so that when the deviations are too great, the physiological oscillations become free-running. The consequences thereof are the theme of this chapter.

Plants. The physiological preference for external cycles that approximately follow the 24-hour periodicity can be demonstrated by the

Table 9. Soybeans, Flowering

LD Cycle	Days until Flowering
hr	
7:7	no flowers
8:9	no flowers
9:9	27
12:12	22
14:14	33
16:16	no flowers
18:18	no flowers

influence of different cycles on the flowering of soybeans. GARNER and ALLARD found the relationship shown in Table 9.

Detailed experiments have been performed by °WENT with LD cycles, or cycles of high and low temperature of various length, in which effects on growth and development of different plants were tested. Cycles close to 24 hours always had optimum effects. The optimum cycle length was somewhat dependent on temperature, and this may be considered to be an expression of the fact that the period of the physiological clock is also slightly dependent on temperature. The correlations may even prove to be somewhat more complicated. KETELLAPPER has reported similar experiments with peanuts and tomatoes. At medium temperatures the optimum length of LD cycles is about 24 hours, at lower temperatures it is a few hours longer, and at higher temperatures it is a few hours shorter.

Flies and humans. PITTENDRIGH and MINIS exposed *Drosophila melanogaster* (adult flies) to three different cycle lengths:

LD 12:12 (24-hour day)
LD 10.5:10.5 (21-hour day)
LD 13.5:13.5 (27-hour day)

The flies raised on 24-hour days lived significantly longer than those raised on false days and longevity of the former was also greater than in the flies exposed to LL.

ASCHOFF et al. found that flies (*Phormia terrae novae*) kept in LD 12:12 hours had an average life of 125 days. However, when simulating transocean flights, the life was reduced to about 98 days. This simulation was by a 6-hour shift of the light-period once a week, in one set simulating eastward flight, in a second set westward flight, in the third set alternating eastward and westward.

Investigations into the problems of air travel and space travel have shown that humans can likewise make only limited adjustment to an environment that deviates considerably from the 24-hour periodicity (see HAUTY and ADAMS; HAUTY et al.; °HAUTY; °HALBERG; °FLAHERTY [ed.]; LaFontaine et al.; *Strughold). The reader is also referred to the interesting experiments of Lewis and Lobban, described in Chapter 4(d), in which humans had access only to clocks that ran fast or slow.

Considerations such as the ones quoted in this section demonstrate the truth of Hufeland's statements (see p. 1), made more than 160 years ago: "That period of 24 hours . . . is, as it were, the unity of our natural chronology."

b. Disturbances by dissociating the rhythms

Plants. The endogenous diurnal rhythm in plants may proceed in some parts of an individual independent of the rhythms of certain other parts. For example, processes may take place with shifted phases in two opposite leaves, if they are exposed to inverse LD cycles. Even within an individual organ, dissociations like these are possible (see Chapter 4(d) and Fig. 28). We can observe the same independence by studying activities bound to different organs. The periodicity of leaf movement is not necessarily synchronized with the periodicity of bleeding and guttation in plants. A disturbance of the normal phase angle difference between the individual rhythms is easily possible if they are not synchronized by the same external factors (Speidel; Engel u. Friederichsen; Heimann). Todt found in *Cichorium* that in LL the rhythm of flower opening becomes desynchronized from the rhythm of pigment formation in the petals.

Animals and humans. In higher animals and humans it is even more obvious that the different parts of an individual do not necessarily show the normal physiological relations in the circadian organization of the body. It has been pointed out that different functions adjust themselves with varying ease to a phase-shifted: LD cycle. Even after a person has worked the night shift for a long period, many body functions retain the original phase position of their rhythm, although the rhythm of sleep has been completely reversed (see Aschoff 1955). This dissociation or "desynchrony" must result in physiological disturbances during the transitional period, i.e., during the days of adjustment, because the rhythm of the different organs normally are phased in such a way that they are advantageous for the orderly cooperation of the organs in an individual. An example of such a physiological disturbance is the behavior of miners on night shift: They "behave as nocturnal animals for potassium excretion and as diurnal animals for the excretion of water" (*Lobban 1965).

We have repeatedly referred to cases in which phases of partial cycles in animals shift with varying ease; for example, the different speed of adjustment in the clock used in orientations, and the rhythm of activity (see pp. 99 and 172).

The experiments of LEWIS and LOBBAN (1957) demonstrate especially well the phase shifts that occur among different diurnal functions in humans under abnormal conditions (*LOBBAN 1960). The test individuals lived at Spitsbergen in the Arctic Ocean. They used watches that made one full turn of the hour hand in 10.5, 11, or 13.5 hours, i.e., they were running fast or slow. The activity rhythm of these persons was correspondingly modified, since a clock naturally has more of a synchronizing influence on conscious activities in humans than does the small difference in light intensity between day and night during the summer at Spitsbergen. As it turned out, the different physiological functions under observation (excretion of water and potassium, body temperature) could be modified with varying ease. The rhythm of potassium excretion, in particular, follows rather strictly the 24-hour periodicity. This quickly results in a disturbance of the normal phase relations among the several periodical functions ("disphasia"). Phase shiftings and dissociations such as these are also evident in the experiments of SHARP at Spitsbergen (p. 64). For further examples on humans, see *LOBBAN (1965) and ASCHOFF (1965).

Dissociations in the absence of LD cycles are not restricted to different physiological functions. Even within the same function as when recording only the locomotor activity, a similar phenomenon may become evident (e.g., Fig. 133; see also HOFFMANN 1971; POHL).

Several authors have described physiological disturbances resulting from journeys by jet planes, such as shiftings in $K^+:Na^+$ ratio, decrease of salt diuresis, etc. Different functions also adjusted themselves with different rapidity to the new time schedule, causing a dissociation of the maxima for the several circadian functions (BUGARD and HENRY; GERRITZEN; HAUTY and ADAMS; LAFONTAINE et al.).

Dissociations such as these, and certainly many additional ones, resulting from flights may be responsible for many well-known consequences of long-distance flights.

> The individual who has crossed a number of time zones, may be in a somewhat handicapped position due to his desynchronic condition. "It has been observed that actors, chess players, athletes, and last but not least, race horses, were not at their best, the first few days after arriving from a region four or more time zones away." (*STRUGHOLD; for further references, *KLEIN et al.)

The phase relations can be disturbed not only by abnormal LD cycles, and in constant conditions, but also by a pathological loss in the normal physiological coupling between the several diurnal func-

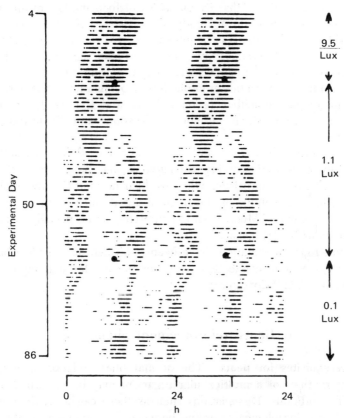

Fig. 133. Locomotor activity of an individual *Tupaia glis* (tree shrew) in LL with different intensities. Black points indicate time of splitting into two distinct components which for some time exhibit different frequencies and later run parallel. Ordinate: days of experiment. After HOFFMANN

tions. We have pointed out the possibility of couplings such as these (see p. 64). Among other things, they are responsible for the fact that after the phase shift of some rhythms in response to a changed LD cycle, the rhythms of organs not directly influenced will also gradually adjust themselves to the shifted phase. If the physiological mechanism of this coupling is impaired, then a change in the normal phase relations will ensue. Phenomena such as these could be a decisive factor in the occurrence of certain diseases.

Provoking certain diseases. Physicians have repeatedly emphasized that disturbed phase relations, regardless of their cause, can result in diseases. FORSGREN has pointed out the possible interference of the rhythms of stomach and liver. Asynchrony among these organs can be the reason for gastric ulcers. It is known, as *MENNINGER-

LERCHENTHAL writes, "that irregularities in food intake are an important factor in the origin of duodenal ulcers . . . No one need doubt this, if he considers that gastric juice is secreted into an empty stomach at noontime, and perhaps only a few hours later the food for which it was secreted comes in contact with it . . ." (translated).

In connection with this example, it may be mentioned that circadian rhythms of gastric acid secretion (MOORE and ENGLERT) and in the level of gastrointestinal hormones (GANGULI and FORRESTER) are known.

If such disturbances last only for a short time, they will result in a generally decreased working capacity, as can be noticed after air flights. This is not surprising, when we bear in mind the investigations cited. More indications of the medical significance of these disturbances are given by *MENNINGER-LERCHENTHAL and especially in several publications by *HALBERG.

Numerous investigations have been published on the disturbances that may occur in connection with working on day and night shifts (*MENZEL 1962; *COLQUHOUN).

c. Beats: reinforcement phenomena

Prerequisites for beats. The normal phase relation among the partial rhythms of a multicellular organism may be disturbed under certain conditions. Dissociations such as these can also occur when there is no physiological coupling between the individual oscillations, or when some of the partial oscillations (e.g., those in individual organs) are not under the control of external synchronizers (e.g., the LD cycles). Coupling usually results either from the influence of one oscillating organ on another or from a common dependence on some controlling centre such as on a hormone secreting gland.

Beats may result, if under such conditions of dissociation, certain circadian oscillations still are entrained to exactly 24 hours by external cues whereas others are free-running with periods deviating from 24 hours. Beats may also result when no synchronization at all is given, but several physiological oscillations differ in the length of the free-running period. A great difference between the inherited and the entrained period may cause beats, too (Fig. 83).

In discussing lunar cycles we found indications of beat phenomena in organisms. Many investigations in humans and animals also support such a possibility. *HALBERG (1960), for example, has pointed this out, quoting as proof some of the results LEWIS and LOBBAN obtained during the Spitsbergen experiments.

Pathological Phenomena

Aschoff (*1965) reports a rather instructive example of a beat phenomenon caused by dissociation under conditions of missing external synchronizers:

An individual living in isolation without time cues exhibited two frequencies even in the same organ, the kidney. The rhythms of activity and calcium excretion regained their original (normal) phase with the rhythms of the three other functions every third to fourth day. The times when all functions were in phase coincided with occasional diary notes in which the subject indicated that he felt especially well and fit.

Periodical diseases. We would also like to draw attention to some phenomena that have long been known to medicine. There are diseases that become manifest only at intervals of several days, weeks, or even longer periods. The comprehensive experimental material gathered by *Richter (see also *Richter) is especially relevant (see also Fig. 134, *Reimann). None of these cases has been analyzed thoroughly enough to decide whether they are the results of reinforcement phenomena, but it is noteworthy in the context of the prerequisites for the beats that they are especially likely to appear in experimental animals when organs with controlling functions are impaired. In rats, Richter found cycle lengths of about 20 days (or even longer) following disorders of the thyroid and pituitary glands. He also described the case of one rat that exhibited such a phenomenon, even though it gave the impression of being healthy at the beginning of the experiment. Later on it turned out that this rat was suffering from a brain tumor which had evidently damaged the glands. Rice has also noticed that cycles of a length of several weeks can originate after the failure of the thyroid or parathyroid gland.

Cases such as these should be investigated to see if they are caused by reinforcements or by lack of coordination among the individually oscillating systems.

*Richter suggested another hypothesis for such rhythms: The

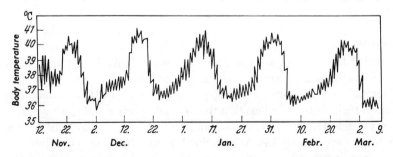

Fig. 134. Daily body temperature of a 19-year-old man with Hodgkin's disease. Peaks are 24–26 days apart. After Ebstein

length of such cycles might correspond to inherent rhythms within the participating cells, or (in other cases) to their lifespan. "After a shock or trauma, the cells may all suddenly be thrown into phase" ("shock-phase hypothesis").

d. Damage due to the absence of synchronizing stimuli

General remarks. We have seen that dissociations will occur when the physiological coordination among the individual rhythms is disturbed and when not all of these rhythms can be synchronized by external stimuli. But dissociations will also occur between two or more different oscillations which are usually synchronized by external stimuli, when these external synchronizers are missing. This will also result in free-running oscillations, which will lead to disturbances.

There is another reason the absence of external synchronizers can result in damage: when the oscillations fade out gradually, the extremes of certain physiological values may no longer be reached. Processes in the cells dependent on these extreme values may become impossible because the necessary prerequisites are no longer provided.

These and other results help to explain the damage which has been observed in plants and animals, especially under the influence of LL. This damage can be eliminated by offering external stimuli from time to time to resynchronize or reinitiate the periodicity. Such a synchronizer, applied perhaps only once a week, is often enough to eliminate such damage.

Damage due to continuous light. Some time ago ARTHUR et al. (1930, 1937) described damage to tomatoes grown under LL. HILLMAN has studied these phenomena in more detail. He noticed that this damage can also be eliminated by applying a diurnal change of temperature. This means that the damage only occurs when neither synchronizing LD cycles nor synchronizing cycles of high and low temperature are available. (See also HIGHKIN; HANSON and HIGHKIN; *WENT).

It is especially interesting that damage caused by LL may be eliminated by simply resynchronizing the rhythms or by reinitiating oscillations that have faded out. For example, some plants do not open their flowers after being exposed to LL for an extended period of time (ARNOLD on *Oenothera,* evening primrose; TODT on *Cichorium*). In *Cichorium* it is quite obvious that the diurnal opening and closing movements of the flowers become more and more desynchronized the longer the time of exposure to LL. At first, this is noticeable among the different inflorescences of one plant, and later it can be observed even among the individual flowers of one inflorescence (see Fig. 28). Concurrent with the more intense desynchronization, the flowers show damage. This damage and the obvious desynchronization or suppression of periodical movements of the petals can be eliminated by a diurnal cycle in light intensity, or by exposing the plants to a single dark period of about 12 hours. Of course, in plants, dark periods of that length generally suffice to reinitiate an oscillation that had faded out in LL (see Chapter 6(b)).

In LL, some strains of the fungus *Pilobolus kleinii* do not form sporangia but only the preceding stage, so-called trophocysts. These accumulate in a culture kept in LL. The block can be released by interrupting the LL with one dark period of about 8 to 10 hours, after which the trophocysts develop into complete sporangiophores. The discharge of the sporangia shows that their formation is controlled by the circadian system (JACOB 1959). This rhythm of sporulation is synchronized by the beginning of the dark period. Here also the elimination of the block evidently means that the dark period reinitiates the rhythm. The physiological extreme values, which had been absent during the period of LL, are reached again. A period of low temperature also has such a releasing effect (JACOB 1961).

Results of experiments with the alga *Oedogonium* (RUDDAT) are quite comparable. This alga bleaches out, i.e., it shows a loss of chlorophyll, in LL. This can be prevented by exposing the cells to a single dark period or to a single period of low temperature (10°C). These interruptions of LL must last for about 6 to 10 hours, but they are not required every day. An interruption of LL by a dark period or a cold period of the same length is also necessary to reinitiate the diurnal sporulation pattern after it has become aperiodic in continuous light.

These experimental findings are strong evidence that a close relationship exists between the reinitiation of a faded-out endogenous periodicity and the elimination of damage due to LL. It remains possible, however, that in some cases the initiation might not really be a reinitiation of faded-out oscillations, but perhaps only resynchronization of several rhythms in a given organism.

References to Chapter 14

(*Pathological Phenomena*)

a. Reviews

*ASCHOFF, J. 1965. Circadian rhythms in man. *Science* **148**:1427–1432.
*COLQUHOUN, W. P. 1971. *Biological Rhythms and Human Performance*. London–New York: Academic Press.

*FLAHERTY, B. E., ed. 1961. *Psychophysiological Aspects of Space Flight.* New York: Columbia University Press.
*GRETER, W. F. 1965. Human performance for military and civilian operations in space. *Ann. N. Y. Acad. Sci.* 134:398–412.
*HALBERG, F. 1960. Temporal coordination of physiologic function. *Cold Spring Harbor Symp. quant. Biol.* 25:289–308;—
——. 1964. Physiologic rhythms. In *Physiological Problems in Space Exploration.* J. D. HARDY, ed. Springfield, Ill.: Charles C Thomas.—
*HAUTY, G. T. 1960. Psychological problems of space flight. In *Physics and Medicine of the Atmosphere and Space.* O. O. BENSON and H. STRUGHOLD, eds. New York: J. Wiley & Sons;—
——. 1962. Periodic desynchronization in humans under outer space conditions. *Ann. N. Y. Acad. Sci.* 98:1116–1125.
*KLEIN, K. E., H. BRÜNER, H. HOLTMANN, H. REHME, J. STOLZE, W. D. STEINHOFF, and H. WEGMANN. 1970. Circadian rhythm of pilot's efficiency and effects of multiple time zone travel. *Aerospace Med.* 41: 125–132.
*LOBBAN, M. C. 1960. The entrainment of circadian rhythms in man. *Cold Spring Harbor Symp. quant. Biol.* 25:325–332;—
——. 1965. Dissociation in human rhythmic functions. In *Circadian Clocks.* J. ASCHOFF, ed. Amsterdam: North Holland Publishing Company. Pp. 219–227.
*MENNINGER-LERCHENTHAL, E. 1960. *Periodizität in der Psychopathologie.* Wien: Maudrich.—
*MENZEL, W. 1962. *Menschliche Tag-Nacht-Rhythmik und Schichtarbeit.* Basel/Stuttgart: Benuo Schwabe & Co.
*REIMANN, H. A. 1963. *Periodic Diseases.* Oxford:Blackwell.—
*RICHTER, C. P. 1965. *Biological Clocks in Medicine and Psychiatry.* Springfield, Ill: Charles C Thomas.
*STRUGHOLD, H. 1965. The physiological clock in aeronautics and astronautics. *Ann. N. Y. Acad. Sci.* 134:413–422.
*WENT, F. W. 1962. Ecological implications of the autonomous 24-hour rhythm in plants. *Ann. N. Y. Acad. Sci.* 98:866–875.

b. Other references

ARNOLD, C. G. 1959. *Planta (Berl.)* 53:198–211.—
ARTHUR, J. M., J. D. GUTHRIE, and J. M. NEWELL. 1930. *Amer. J. Bot* 17:416–482.—
ARTHUR, J. M., and E. K. HARVILL. 1937. *Contr. Boyce Thompson Inst.* 8:433–443.—
ASCHOFF, J. 1955. *Naturwiss.* 42:569–575.—
ASCHOFF, J., U. v. SAINT PAUL, u. R. WEVER. 1971. *Naturwiss.* 58:574.
BUGARD, P., et M. HENRY. 1961. *La Presse Médicale No.* 44:1903.
EBSTEIN, W. 1887. *Berl. klin. Wschr.* 24:565.—
ENGEL, H., u. I. FRIEDERICHSEN. 1952. *Planta (Berl.)* 40:529–549.

FORSGREN, E. 1947. *Acta med. scand.* **128**:281–288;–
———. 1947. *Nord. med. T.* **34**:1280.
GANGULI, P. C., and J. M. FORRESTER. 1927. *Nature, New Biology* **236**:127–128.–
GARNER, W. W., and H. A. ALLARD. 1920. *J. agr. Res.* **18**:553–606.–
GERRITZEN, F. 1962. *Aerospace Medicine* **33**:697–701.–
GHATA, J. 1967. *Aerospace Medicine* **38**:944–947.
HANSON, J. B., and H. R. HIGHKIN. 1954. *Plant Physiol.* **29**:301–302.–
HAUTY, G. T. 1963. *Aerospace Medicine* **34**:100–105.–
HAUTY, G. T., and T. ADAMS. 1966. *Aerospace Medicine* **37**:1027–1033.–
HAUTY, G. T., G. R. STEINKAMP, W. R. HAWKINS, and F. HALBERG. 1960. *Fed. Proc.* **19**:54.–
HEIMANN, M. 1950. *Planta (Berl.)* **38**:157–195.–
HIGHKIN, H. R. 1960. *Cold Spring Harbor Symp. quant. Biol.* **25**:231–238.–
HILLMAN, W. S. 1956. *Amer. J. Bot.* **43**:89–96.–
HOFFMANN, K. 1969. *Zool. Anz. Suppl. Bd. 33, Verh. Zool. Ges.* 171–177;–
———. 1971. In *Biochronometry*. M. MENAKER, ed. Pp. 134–151. Washington, D.C.: Nat. Acad. Sci.
JACOB, F. 1959. *Arch. Mikrobiol.* **33**:83–104.–
———. 1961. *Flora* **151**:329–344.
KETELLAPPER, H. J. 1960. *Plant Physiol.* **35**:238–241.
LAFONTAINE, E., J. LAVERNHE, J. COURILLON, M. MEDVEDEFF, and J. GHATA. 1967. *Aerospace Medicine* **38**:944–947.–
LEWIS, P. R., and M. C. LOBBAN. 1957. *J. exp. Physiol.* **42**:371–386.
MOORE, J. G., and E. ENGLERT. 1970. *Nature* **226**:1261–1262.
PITTENDRIGH, C. S., and D. H. MINIS. 1972. *Proc. Nat. Acad. Sci. USA* **69**:1537–1539.–
POHL, H. 1972. *J. comp. Physiol.* **78**:60–74.
RICE, K. 1944. *Arch. Neurol. Psychiat. (Chic.)* **51**.–
RICHTER, C. P. 1957. *Recent Progr. Hormone Res.* **13**:105–159;–
———. 1960. *Proc. Nat. Acad. Sci. USA* **46**:1506–1530.–
RUDDAT, M. 1960. *Z. Bot.* **49**:23–46.
SPEIDEL, B. 1939. *Planta (Berl.)* **30**:67–112.–
STRUGHOLD, H. 1962. *Ann. N. Y. Acad. Sci.* **98**:1109–1115.
TODT, D. 1962. *Z. Bot.* **50**:1–21.

Author Index

Numbers in *italics* indicate the pages on which the references are listed.

Aaker, H., 68
Abraham, G. H., 39, *47*
Abramowitz, A. A., 52, *66*
Adams, T., 100, *114*, 230, 232, *239*
Adkings, G., 5
Adkisson, P. L., 199–201, 205, 215, 217, 219, 221, *223*, *225*, 228
Adler, K., 62, *66*, 98, *112*
Admundson, S. J., 222, *224*
Agren, G., 57, *66*
Albe, D., 128, *133*
Albertus Magnus, 196
Albrecht, P. G., *113*
Allard, H. A., 198, *226*, 230, *239*
Allen, C. F., 61, *68*
Altukhov, G., 164, *165*
Andrews, R. V., 50, *66*, 75, *87*
Androsthenes, 14
Ankersmit, G. W., 200, 201, 207, *225*
Aranow, R. H., 151, *153*
Arnold, C. G., 236, *238*
Arthur, J. M., 236, *238*
Arvanitaki, A., 128, *133*
Aschoff, J., 5, 11, 12, 16, 19, 20, 22, 23, 28, 29, 43, *46*, 47, 56, 63, *65*, 77, *86*, 89, 90, 93, 94, 100, 104, *112*, 159, 163–165, 199, 207, 215, *223*, *223*, 225, 230–232, 235, *237*
Assenmacher, I., 61, *65*
Atchley, F. O., *66*
Auger, D., 128, *133*
Austin, B., 8, *29*
Autrum, J., 167, *174*
Axelrod, J., 62, *66*, 69, 70, *115*
Azarjan, A. G., 54, *66*

Badino, G., 173, *175*
Badran, A. M. F., 49, *66*
Bahorsky, M. S., 62, *66*
Baker, B. L., 60, *67*
Baker, I. Z., 201, *225*
Baker, J. R., 201, *225*
Bakke, J. L., 60, *66*
Baldaccini, N. E., *176*
Ball, N. G., 18, *29*, 34, 41, *46*, 80, *87*, 138, 139, *153*
Baltes, J., 140, 141, *153*
Baranetzky, J., 15, *29*
Barker, R. J., 207, *225*
Barnett, A., 48, *66*, 145, *153*
Barnum, C. P., 38, *46*, 65, *68*, *113*, *114*, 145, 152, *154*
Barnwell, F. H., 177, *187*
Beane, W. L., 228
Beck, S. D., 199, 215, *223*
Becker, G., 26, *29*
Beier, W., 161, *165*
Beisel, W. R., *67*
Belai, V. E., *165*
Beling, I., 2, 6, 16, *29*, 160, *165*
Belleville, R. E., *174*
Bennett, J. W., 199, *225*
Bennett, M. F., 10, *29*, 67, *165*, 177, *187*, *226*
Benoit, J., 61, *65*, 199, 215, 217, *223*
Bentley, E. W., 80, *87*
Benzer, S., 24, *31*
Berger, S., *155*
Berliner, M. D., 14, *30*
Bernardis, L. L., 62, *66*
Bernhardt, W., 141, *153*
Berry, C. E., *165*
Betz, A., 141, *153*

Bianchi, D. E., 13, *30*
Biebl, R., 144, *153*, 154, 192, *194*
Bierhuizen, J. F., *5*
Bindon, B. M., 61, *66*
Binkley, S., 62, *66*
Birukow, G., 169, 171–173, *174–176*, 183, *187*, 190, *194*
Bissonette, Th. H., 217, *225*
Bittner, J. J., *46*, 65, 68, *113*, *114*, 152, *194*
Blalock, 63
Blanley, L. T., 213, *225*
Bliss, D. E., 52, *66*
Blume, J., 26, *30*, 110, *113*, 129, *133*, 179, *187*
Bode, V. C., 138, *153*
Böttcher, H., *174*
Bogorov, B. G., 25, *30*
Bohn, G., 147, *187*
Bonga, H., 204, *226*
Bonotto, S., 146, *156*
Borthwick, H. A., *223*, *226*, *228*
Bose, 125
Boughton, D. C., 62, *66*
Bourret, J. A., 14, *30*
Boyd, G., 62, *67*
Bracher, R., 176, *187*
Brachet, J., *156*
Brady, J., 53–55, *65*, *67*
Braemer, W., 169, 172, 173, *174–176*
Bramwell, 19
Brehm, E., 100, *113*
Bremer, H., 17, *30*
Brest, D. E., 212, 214, *225*
Brett, W. J., 36, *46*
Bretzl, H., 14, *30*
Briggs, W. R., 32, 99, *115*
Brinkmann, K., 48, *67*, 73, *87*, 139, 142, 149, *153*
Browman, L. G., 22, *30*
Brown, F. A., *5*, 26, *30*, 35, *46*, 52, *67*, 80, *87*, 89, 101, 109, *152*, 163, 164, *164–166*, 177, 185, *187*
Brown, H. E., 61, *67*
Brown, W., *194*
Bruce, V. G., 14, 24, *28–30*, 35, 41, *46*, 48, *67*, *69*, 72, 77, 80, *87*, 88, 89, 99, 101, *113–115*, 124, *133*, 138, 140, *153*, 215, *225*, *228*
Brun, R., 167, *175*
Brüner, H., *114*, *238*

Brunken, W., 89, *115*
Brunt, E. E. V., 98, *113*
Buck, J. B., 17, *30*
Bugard, P., 232
Bühler, A., 13, *30*
Bühnemann, F., 8, 18, *30*, 74, 79, 80, *87*, 136, 137, *153*
Bünning, E., 8, 10, 13, 15–18, 21–23, 27, *28*, *30*, *32*, 34, 36, 37, *46*, 49, 50, *67*, 71, 73–76, 79, 80, 82–86, *87*, *88*, 95, 96, 104, 109, 110, *113*, 121, 125, 127–131, *133*, *134*, 138, 140, 141, 143, 147, 150, 152, *153*, *154*, 184, 185, *187*, 198, 202, 204, 206, 209, 212–215, 217, 221, *223–225*
Bünsow, R., 8, *30*, 208, 212, 215, *224*, *226*
Burger, J. W., 217, *226*
Burns, E. R., 193, *194*
Burton, A. C., 93, *113*
Busch, E., 171, *175*
Büsch, G., 147, *154*
Buttrose, M. S., 147, *154*

Cain, J. R., 89, *113*
Calhoun, J. B., 80, *87*
Cameron, L. L., 13, *28*, 29
Candolle, De, 34
Carpenter, B. H., *30*, *165*, 215, 216, *226*
Caspers, H., 17, *28*, 182, *186*, *187*
Chance, B., 141, *153*, *154*
Chandrashekaran, M. K., 96–98, *113*, *114*, 177, *187*
Chaudhri, I. I., 222, *226*
Chedid, A., 149, *154*
Chou, Ch., 70
Chovnick, A., *5*
Christie, A. O., 183, *187*
Chu, E. W., 70
Cihlar, J., 127, 128, *134*
Claes, H., 208, *226*
Claret, J., 218, *226*
Clark, R. H., 60, *67*
Clauser, G., 3, *6*, 17, 19, *30*, 162, *164*
Clauss, H., 10, *30*, 146, 147, *154*
Clegg, M. T., *113*
Clements, H. F., 221, *226*

Author Index

Cloudsley-Thompson, J. L., 5, 98, 113, 157, 158, 163, *164*, *165*, 182, *186*
Cohen, W. D., 135, *154*
Collins, A., *156*
Colquhoun, W. P., 11, *28*, *30*, *32*, *234*, *237*
Conroy, R. T. W. L., 5
Cook, J., 146, 154
Corbet, P. S., *187*
Coulter, M. W., 207, 226
Courillon, J., *114*, *239*
Cox, V. J., 8, *47*
Crescitelli, F., 190, *194*
Crombie, A. C., 196
Cumming, G., 152, *156*, 211, 213, *224*, *226*, 227
Cymborowski, B., 52, 53, 67

Dainton, B. H., 79, *87*
Damaschke, K., 26, 29
Dangerfield, H. G., 67
Danilevskii, A. S. (identical with Danilevsky), 23, 24, *30*, 199–201, 217, *224*
Darin, D. L., *87*
Darwin, Ch., 21, 121, *134*
Darwin, F., 21, *134*
Dassanayake, W. L., 67
Davenport, D., 189, *194*
Davis, K. B., 68
Davis, W. M., 191, *194*
DeCoursey, P. J., 18–20, *30*, 103, *113*
De Luca, H. F., *156*
De Silva, J., 67
Demmelmeyer, H., 100, *113*
De Mairan, M., 7, 15, *30*
Demoll, R. M., 7, 15, *30*
De Wilde, J., 199, 204, 217, *226*
Denney, A., 211, 228
Dessauer, H. C., 220, 222, *226*
Dietlein, L. F., *165*
Diffley, M., 5
Di Mila, A., 177, *187*
Dolnik, V. R., 222, *226*
Donachie, W. D., 141, *155*
Dowse, H. B., 140, *155*, 163, *165*
Driskill, R. J., 69
Dumortier, B., 54, 67, 98, *113*
Durkee, F., 32

Dyke, I. J., 29, 34, 41, *46*, 80, 87, 138, 139, *153*

Ebbecke, U., 38, *46*
Ebstein, W., 235
Edmunds, L. N., 14, *30*, 135, 145, *154*, 155
Egorov, A. D., *165*
Ehrenberg, M., 135, *154*
Ehrengut Lange, J., 28
Ehret, Ch., 48, 67, 99, *113*, 146, 148, *154*, 158, *165*
Eichhorn, J., 228
Eidmann, H., 53, 67
Elliot, J., 227
Elsaesser, S., *154*
Emeis, D., 82, 87, 171, 173, *175*
Emlen, S. T., 191, *194*
Enderle, W., 50, 67
Engel, H., 231, 238
Engel, R., 63, *66*, 67
Engeli, M., *31*
Engelmann, Th. G., 44, 45, *47*
Engelmann, W., *47*, 96, 98, 106, *113*, 152, *154*
Engels, W. L., 206, 207, *227*
Englert, E., 234, *239*
Enright, J. T., 80, *87*, 107, *112*, 140, 151, *154*, 164, *165*, 174, *175*, 180–182, 184, 186, *187*
Ercolini, A., 172–174, *175*, 176
Erikson, L.-O., 98, *113*
Erkert, H. G., 95, *113*
Erkinaro, E., 110, *113*
Ertel, R. J., 192, *194*
Eskin, A., 18, *30*, 51, 67, 152, *154*, 163, *166*
Eskridge, L. C., *66*
Estabrook, R. W., *154*
Euler, U. S. v., 57, 67
Evans, L. T., 211, 216, *224*
Evans, L. V., 183, *187*
Everett, J. W., 60, 67
Ewer, D. W., 87

Farner, D. S., 199, 207, 215, 219, 222, *224*, *226*
Fatranska, M., *165*
Fauré-Fremiet, E., 177, 178

Feigin, R. D., 62, 67
Feldman, J. F., 134, *154*
Ferguson, D. D., 60, 64, 67, 68
Ferguson, D. E., 98, *115*, 169, *175*
Ferguson, R. B., *154*
Fessard, A., 128, *134*
Fiaschi, V., *176*
Fingerman, M., 52, 54, 67, 177, *186*, *187*
Finlayson, L. H., 53, *69*
Finn, J. C., *165*, 213, *226*
Fiore, L., *176*
Fischer, A., 51, 67
Fischer, H., 144, *154*
Fischer, J. E., *69*, *70*
Fischer, K., 169, 173, *174*, *175*, 220, *226*
Flaherty, B. E., 231, *238*
Flügel, A., 101, *113*
Folk, G. E., 50, *66*, 75, 87
Forel, A., 16, *30*, 159, *165*
Forrester, J. M., 234, *238*
Forsgren, E., 16, 17, *30*, 233, *239*
Forward, E. F., 189, *194*
Fowler, D. J., 50, 67
Fox, S., 227
Fox, W., 220, 222, *226*
Franck, G., 68
Frank, K. D., 99, *113*
Fraps, R. M., 60, 61, *65*, 67
Frazer, J. F. D., 222, *224*
French, L. A., 68
Friederichsen, I., 231, *238*
Frisch, K. v., 160, 167, 170, 175
Funch, R. R., 145, *154*
Funck, H., 17, *31*
Furuya, S., 50, 67

Galicich, J. H., 58, 60, 61, 67, 68
Galler, S. R., 167, 169, 174
Galston, A. W., 152, *155*
Gambino, J. J., *194*
Ganguli, P. C., 234, *239*
Ganong, W. F., *113*
Garner, W. W., 198, *226*, 230, *239*
Gaston, L. K., *166*
Gaston, S., 62, 68
Geisler, M., 21, *31*, 92, 100, *113*
Geppetti, L., 177, *187*
Geronimo, J., 142, *154*
Gerritzen, F., 17, *31*, 232, *239*

Geyspitz, K. F., 217, *226*
Ghata, J., 5, *32*, *114*, *239*
Ghosh, A., *154*
Gibbs, S. J., 146, *154*
Gibson, R. N., 179, 182, *187*, *194*
Giedke, H., *165*
Giersberg, H., 60, *69*
Ginet, R., 26, *31*
Glick, D., 58, 68, 135, *154*
Glick, J. L., 135, *154*
Goodnight, C. J., 50, 67
Goodwin, B. C., 27, *32*, 141, *152*
Goryshin, N. I., *224*
Grabensberger, W., 138, *154*
Grassé, M. P., 217, *226*
Grassi, M., 170, *176*
Graven, H., *68*
Greenberg, L. J., *68*, *154*
Greter, W. F., *238*
Gropp, A., 49, *68*
Guilford, C. B., 10, *29*
Günzler, E., 26, *30*, *31*
Gurevitch, B. Kh., 51, *68*
Guthrie, J. D., *238*
Guttenberg, H. v., 126, *134*
Gwinner, E., 163, *165*, 214, 222, *226*

Haarhaus, D., 110, *113*
Hague, E. B., 5, 61, *65*
Hailer, G., 87, *113*
Halaban, R., 90, 99, *113*, 212, 214, *226*
Halberg, E., *65*, *114*, *152*
Halberg, F., 1, 5, 20, *31*, *32*, 38, *46*, 50, 57–60, *65*, 67–70, 97, 100, *113*, *114*, 135, 144–146, *152*–*154*, 164, *165*, 189, 191–193, *193*, *194*, 230, 234, *238*, *239*
Hammond, J., 206, 226
Hamner, K. C., 39, 47, 163, *165*, 207, 212, 213, 215, 216, *224*, *225*, *226*
Hamner, W. M., 207, 213, 215, *226*
Handler, S. D., 135, *155*
Hanson, J. B., 236, *239*
Hanström, B., 52, *68*
Hardeland, R., 135, *154*
Harker, J., 43, *46*, 52, 53, *65*
Harner, R., 68
Harris, S. J., *32*
Hart, D. S., 199, 205, *226*
Hartland-Rowe, R., *187*

Hartmann, K. M., 48, 68
Harvey, R., 8, 32
Harvill, E. K., 238
Hasler, A. D., 169, 173, 174–176
Hassbargen, H., 94, 114
Hastings, J. W., 5, 10, 11, 28, 72, 73, 80, 87, 88, 101, 114, 132, 134, 135, 136, 138, 139, 142, 144–146, 148, 149, 152–155
Hauenschild C., 182, 183, 186, 187
Haus, E., 191, 193, 194
Hauty, G. T., 100, 114, 230, 232, 238, 239
Hawking, F., 63, 65, 68
Hawkins, W. R., 239
Haxo, F. T., 145, 155
Hayden, P., 194
Heath, W. G., 191, 194
Heimann, M., 231, 239
Hellbrügge, T., 22, 28, 44, 46, 47
Hemmingsen, A. M., 16, 22, 31, 89, 92, 114
Hempel, G., 24, 31, 100, 113
Hempel, I., 24, 31
Hendricks, S. B., 216, 226–228
Henfrey, A., 196, 227
Henry, M., 232, 238
Hess, C., 147, 154
Highkin, 236, 239
Hill, J., 92, 114
Hill, J. C., 201, 228
Hill, M., 61, 68
Hillman, D., 31
Hillman, W. S., 197, 211, 214, 215, 226, 227, 236, 239
Hinds, D. S., 31
Hines, M. N., 187
Hissen, W., 66
Hofer, K., 144, 153, 154, 192, 194
Hoffman, F. M., 147, 154
Hoffmann, K., 19, 22, 31, 64, 68, 73, 79, 80, 87, 89, 90, 112, 167, 168, 171–174, 174, 175, 232, 233, 239
Holdsworth, M. B., 99, 114
Holick, M. F., 156
Hollwich, F., 217, 224
Holmquist, A. G., 57, 67
Holowinsky, A. W., 49, 70
Holtmann, H., 114, 238
Holzer, H., 153
Honegger, H.-W., 98, 113, 179, 182, 187, 188

Honma, K., 20, 29
Horstmann, C., 22, 31
Hoshizaki, T., 165, 225
Howes, C. E., 228
Hoyt, W. D., 183, 187
Hufeland, C. W., 1, 15, 229, 231
Hupe, K., 49, 68
Hutchinson, J. C. D., 225, 226

Ingold, C. T., 8, 47
Ioffe, A. A., 51, 68
Isaac, I., 39, 47
Isaak, D., 194
Isch, F., 134
Ives, D., 98, 115
Izquierdo, J. N., 146, 154

Jacklett, J. W., 51, 68, 142, 154, 190, 194
Jacob, F., 237, 239
Jacobs, G. J., 174
Jahn, T. J., 190, 194
Janda, V., 52, 68
Jander, R., 173, 175
Jegla, Th. C., 26, 31
Jenner, Ch. E., 201, 202, 204, 206, 207, 227, 228
Jerebzoff, S., 14, 28, 31
Jigajinni, S. G., 99, 114
Joerrens, G., 202, 206, 213, 215, 217, 226
Johnson, E., 194
Johnson, M., 89, 90, 114
Johnson, M. S., 16, 31
Johnsson, A., 132, 134
Jones, D. A., 179, 182, 187
Jones, M. B., 10, 31
Jores, A., 17, 28, 56, 68, 192, 193
Jorpes, E., 66
Junges, W., 197, 227
Justice, K. E., 31

Kaiser, I. H., 189, 194
Kalmus, H., 16, 17, 28, 31, 52, 59, 60, 68, 71, 72, 80, 87, 138, 154, 172, 175
Kandeler, R., 197, 227
Kannwischer, L. R., 61, 68
Karakashian, M. W., 12, 31, 48, 68, 145, 148, 154, 158, 165

Karlsson, H. G., 132, *134*
Karve, A., 41, *47*, 99, *114*
Katayama, T. C., 200, 201, 204, 227
Katchalsky, A., 151, *153*
Kaus, P., *115*
Kauzmann, W., 142, *155*
Kayser, Ch., 36, 222, *224*
Keats, H., *227*
Keeton, W. T., 168, *175*
Keil, L. C., *66*
Keil, N. N., *66*
Keller, J. G., 22, *32*
Keller, S., 139, 140, *155*
Kem, W., 192, *194*
Kendall, J. W., 61, *68*
Kerling, L. C. P., 201, *227*
Ketellapper, H. J., 157, *166*, 230, *239*
Keynan, A., *28*, 132, *134*, 146, 148, *153*
Kian Eng Chua, *194*
Kiesel, A., 16, *31*
King, R. W., 211, *227*
Kittlick, P.-D., 152, *155*
Klapow, L. A., 180, 182, *187*
Kleber, E., 159, 160, *165*
Klebs, G., 196, *227*
Klein, K. E., 93, *114*, 232, *238*
Kleinholz, L. H., 51, 52, *68*
Kleinhoonte, A., 9, 15, 21, *31*, 101, 106–108, *114*
Kleitman, E., 55, *68*
Kleitman, N., 5, 9, 17, *28*, *31*, 44, 45, *47*, 55, *68*, 107, 109, *114*
Klitzing, L. v., 143, *155*
Klotter, K., 12, *28*, *86*, 116, *133*
Klug, H., 53, *68*, 144, *155*
Kluth, E., *66*
Knight-Jones, E. W., 183, *187*
Kogure, M., 199, 200, *227*
Koller, G., 51, *68*
Koltermann, R., 160, *165*
Könitz, W., 35, 92
Konopka, R. J., 24, *31*
Korringa, P., 186, *187*
Kosichenko, L. P., 145, *156*
Kramer, G., *114*, 167, 168, *175*
Krarup, N. B., 16, 22, *31*, 89, 92, *114*
Krehan, I., 219, *227*
Kristoffersen, T., 157, *165*
Kübler, F., 51, *68*

Kurras, S., 133, *153*
Kytomäki, O., 58, *69*

Lafontaine, E., 93, *114*, 231, 232, *239*
Lago, A. D., *67*
Lamond, D. R., 61, *66*
Lamprecht, G., 107, 110, *114*
Landreth, H. F., *175*
Lang, A., 208, 213, *224*, 227
Lang, H. J., *187*, 226
Lange, J., *28*, 47
Lange, O. L., 190, 191, *195*
Lardner, P. J., *31*
Larimer, J. L., 98, *114*
Lavernhe, J., *114*, 239
Lawrence, N., 60, *66*
Le Bouton, A. V., 135, *155*
Lee, R., *68*
Lee Kavanau, J., 109, *114*
Lees, A. D., 199–202, 207, 208, 215, 217, 218, *224*, 227
Leighton, A. T., 219, *227*
Leinweber, F. J., 18, *31*, 73, 74, *87*
Levengood, M. C., 10, *31*
Levy, L. M., *67*
Lewis, P. R., 55, *68*, 231–234, *239*
Lewis, R. A., 199, 222, *224*
Lincoln, R. G., 30
Lindauer, M., 161, *165*, 166, 172, *175*
Lindberg, R. G., 192, *194*
Lipton, G. R., 27, *31*
Lisk, R. D., 61, *68*
Livingston, L., *32*
Llanos, J. M. E., 49, *66*
Lobban, M. C., 25, *31*, 55, *68*, 231, 232, 234, *238*, *239*
Lofts, B., 197, 199, 222, *224*, 227
Lohmann, M., 18, *31*, 132, *134*
Loosanoff, V. L., 183, *187*
Lörcher, L., 40, *47*, 90, 91, 99, *114*
Lowe, Ch. H., 18, *31*
Lowe, M. F., *67*
Lowenstein, W. R., 152, *155*
Lowry, R. W., *32*
Luce, G. G., 5
Lüters, W., 169, *176*
Lutsch, E. P., 192, *194*
Lutz, F. E., 17, *31*
Lyman, H., 146, *155*

Author Index

MacDowall, F. D. H., 148, 155
Macey, E. J., 165
Mansfield, T. A., 10, 31
Marcovitch, S., 199, 227
Marshall, A. J., 197, 219, 222, 224, 227
Martinez, J. L., 89, 114
Marx, Ch. H., 128, 134
Masters, S., 141, 155
Matthews, J., 68, 167, 168
Matthies, E., 194
Mauleon, P., 228
Mayer, W., 25, 31, 76, 80, 87, 113, 114
Mayersbach, H. v., 149, 153
McDonald, D. L., 168, 176
McDowall, F. D. H., 148, 155
McKeown, J. P., 175
McLeod, D. G. R., 162, 165
McMillan, J. R., 62, 68, 98, 114
McMurry, L., 142, 149, 155
McWilliams, P., 102, 104
Medugorac, L., 161, 165, 166
Medvedeff, M., 114, 239
Meir, A. H., 60, 68, 193, 194
Melby, J., 68
Melchers, G., 210, 227
Menaker, M., 5, 19, 30, 31, 62, 66, 68, 69, 87, 96, 98, 114, 163, 166, 207, 214, 215, 222, 224, 227
Meng, K., 192, 194
Menninger-Lerchenthal, E., 193, 193, 233, 234, 238
Menzel, W., 56, 65, 193, 194, 234, 238
Mercer, D. M. S., 116, 133
Merkel, F. W., 222, 227
Merrit, J. H., 146, 155
Mewaldt, L. R., 207, 219, 226
Meyer, A., 163, 166
Meyer, R. K., 61, 69
Meyer-Lohmann, J., 22, 29, 43, 89, 114
Michener, M. C., 169, 176
Miller, A. H., 219, 227
Miller, J. H., 147, 154
Millet, B., 42, 47
Mills, J. N., 5
Minis, D. H., 36, 47, 99, 113, 215, 225, 227, 228, 230, 239
Miselis, R., 169, 176

Mitchison, J. M., 14, 29
Mitrakos, K., 146, 155
Miyata, H., 10, 31
Mödlinger-Odorfer, M., 144, 155
Moore, J. G., 234, 239
Morgan, E., 177, 182, 187
Mori, S., 31, 109, 114
Morris, L. R., 199, 227
Morris, R. W., 192, 194
Morris, T. R., 220, 227
Moser, I., 77, 78, 87, 102–104, 113, 114, 130, 131, 133, 150, 153, 204, 214, 226
Mottram, J. C., 192, 194
Müller, D., 184, 185, 187, 196, 227
Müller, H. J., 217, 227
Müller, K., 110, 114
Müller, M., 60, 69

Naegele, J. A., 194
Nagarcenkar, R., 220, 227
Nair, V., 149, 154
Naylor, E., 52, 69, 86, 88, 179, 181, 182, 185, 187, 188
Nemec, S. J., 183, 188
Neumann, D., 179, 184, 187, 188
Neurath, P. W., 14, 30
Neuscheler, W., 189, 194
Newell, J. M., 238
Niebroj, T., 144, 155
Nishiisusuji-Kwo, J., 53, 54, 69
Nishimura, M., 154
Njoku, E., 201, 227
Noles, P. K., 219, 228
Nomejko, C. A., 183, 187
Norris, M. J., 219, 227
Nowosielski, W., 21, 31, 98, 114, 194
Nunnely, S. A., 165

Oatley, K., 27, 32
Oberdorfer, H., 172, 175
Ogasawara, F. X., 219, 227
Oguro, Ch., 52, 67
Oltmanns, O., 75, 79, 88
Omdahl, J., 156
Opel, H., 61, 69
Ortavant, R., 197, 228
Overland, L., 159, 166

Padilla, G. M., 13, *28*, 29
Page, T. L., 98, *114*
Palmer, J. D., 5, 11, 18, *32*, 140, *152*, *155*, 163, *165*, 179, *188*, 190, *194*
Panofsky, H., 5
Panten, K., *153*
Papi, F., 80, *80*, 169–174, *176*
Pardi, L., 169–174, *176*
Paris, O. H., 201, 202, 204, *228*
Park, O., 17, 22, 26, *29*, *32*
Park, Y. H., *165*
Parker, M. W., *228*
Parkes, A. S., 61, *68*
Parrini, S., 88
Parthier, B., 146, *153*
Patton, R. L., 21, *31*, 98, *114*
Pauly, J. E., 59, 60, *69*, 146, *155*, *194*
Pavlidis, T., 142, *155*
Payne, L. F., 217, *228*
Peiponen, V. A., 25, *32*
Pengelley, E. T., 222, *225*
Petren, T., 44, *47*
Petrovic, V., 59, *69*
Pfeffer, W., 13, 15, *32*, 110, *114*
Pincus, G., 57, *69*
Pirson, A., 135, *155*
Pittendrigh, C. S., 18, 25, *29*, *32*, 35, 36, *46*, *47*, 48, 53, 54, 64, 67, *69*, 72, 77, *87*, 88, 89, 93, 101, 102, 109, 110, *112*, 113–115, 120, 124, *133*, *134*, 138, 140, *153*, 215, *225*, *228*, *230*, *239*
Pitts, G. C., *165*
Pizzarello, D. J., 192, *194*
Pohl, H., 36, *47*, 64, *69*, 88, *115*, 232, *239*
Pohl, R., 48, *69*
Pohle, K., 192, *194*
Polcik, B., 192, *194*
Popovic, P., 59, *69*
Poppel, E., 163, *166*
Poulson, Th. L., 26, *31*
Pressman, B. C., 150, *155*

Quay, W. B., 61, 62, *69*, 144, *155*
Quinke, H., 17, *32*

Rakha, A. M., 61, *69*
Ralph, C. L., *113*, *165*
Rao, K. P., 177, 181, *188*
Rau, W., 146, 147, *154*

Rawson, K. S., 75, 88, 140, *155*, 172, *176*
Rehme, H., *114*, *238*
Reimann, H. A., 235, *238*
Reinberg, A., 5, 20, *31*, *32*, 192, *194*
Remmert, H., 9, *29*, 109, *115*, 157, 158, *164*, 190, *194*
Renner, M., 80, 81, 88, 138, *155*, 161, *164*
Rensing, L., 10, 22, *29*, *32*, 43, *46*, 51, 52, *69*, 82, 88, 89, *115*, 193, *194*
Renzoni, A., 144, *155*
Resende, F., 219, *228*
Retiene, K., 61, *65*
Rice, K., 235, *239*
Richter, C. P., 5, 13, 16, *29*, *32*, 59, 60, 62, *66*, *69*, 235, *238*, *239*
Richter, G., 135, *155*
Riddiford, L. M., 54, 55, *69*
Rimet, M., 189, *194*
Ringoen, A. R., 217, *228*
Rinne, U. K., 58, *69*
Roach, S. H., 219, 221, *225*, *228*
Roberts, A. M., 163, *166*
Roberts, R., 227
Roberts, S. K., 18, *32*, 52–54, 60, *66*, *69*, 79, 82, 84, 88, 98, 105, 109, *115*
Robertson, H. A., 61, *69*
Rodriguez, G., 182, *188*
Roff, M. F., 162, *165*
Rohles, F. H., *6*
Rohmer, F., *134*
Rose, B., 152, *155*
Rose, Ch. M., *70*
Röseler, I., 51, *69*
Rowan, W., 199, *228*
Rubin, R. H., *155*
Ruddat, M., 39, 43, *47*, 82, 88, 237
Rummel, J. A., *165*
Rutenfranz, J., *28*, *47*

Sachs, J., 15, *32*, 89
Sachs, R. M., 221, *228*
Saint Girons, M. C., 36, *47*
Salisbury, F. B., 211, *228*
Sandeen, M. I., *187*
Sargent, M. L., 14, *32*, 99, *115*
Satter, R. L., 152, *155*
Saunders, D. S., 207, *228*
Sawyer, Ch. H., 60, *67*

Author Index

Schäfer, E. A., 199, 228
Scherer, L. E., 36, 47
Scheving, L. E., 59, 69, 146, 155, 192, 194
Schmidle, A., 8, 32
Schmidt-Koenig, K., 100, 115, 168, 171, 174, 176
Schnabel, G., 108, 109, 115
Schölm, H. E., 144, 155
Schöne-Schneiderhöhn, G., 143, 154
Schöner, B., 154
Schoser, G., 47
Schrank, F. v. P., 116
Schwassmann, H. O., 169, 172, 173, 174, 176
Schweiger, E., 146, 155
Schweiger, H. G., 145, 146, 155
Schwemmle, B., 10, 30, 146, 154, 157, 164, 190, 191, 195
Scott, H. M., 217, 228
Scully, N. J., 228
Seaman, G. V. S., 62, 66
Sel'kov, E. E., 6
Semon, R., 15, 32
Serretti, L., 88, 176
Serventy, D. L., 222, 227
Sestan, N., 144, 155
Shah, V. C., 146, 155
Shank, M. C., 222, 228
Sharp, G. W. G., 64, 69, 93, 99, 115, 232
Shepherd, M. D., 113
Shoffner, R. N., 219, 227
Shorey, H. H., 166
Shriner, J., 113
Siegel, H. S., 219, 228
Siffre, S., 31, 32
Sirohi, G. S., 165
Skopic, S. D., 36, 47, 69
Smith, D., 113
Smith, R. E., 219, 228
Snyder, S. H., 61, 69, 115
Sollberger, A., 6, 116, 133
Sonneborn, T. M., 48, 69, 158, 166
Sower, L. L., 158, 166
Spalding, J. F., 192, 194
Spangler, R., 151, 153
Speidel, B., 231, 239
Stadler, D. R., 24, 32
Stanners, C. P., 192, 194
Steht, K., 28
Stein, M., 5

Steinheil, W., 140, 155
Steinhoff, W. D., 114, 238
Steinkamp, G. R., 239
Stephan, F., 61, 69
Stephen, W. P., 10, 32, 93, 99, 115
Stephens, G. C., 82, 88
Stern, K., 15, 30, 79, 80, 88
Stolze, J., 114, 238
Stoppel, R., 15, 32
Strauss, W. F., 61, 69
Stross, R. G., 201, 228
Strughold, H., 164, 166, 231, 232, 239
Strumwasser, F., 50, 69, 183, 231
Sulkowski, T. S., 146, 155
Sulzman, F. M., 135, 155
Suter, R. B., 140, 155
Sutherland, D. J., 27, 31
Sussman, A. S., 14, 32
Suzuki, N., 31
Swade, R. H., 25, 32, 109, 110, 115
Swank, R. L., 66
Sweeney, B. M., 6, 10, 11, 28, 29, 39, 47, 72, 73, 80, 87, 88, 99, 101, 114, 115, 135, 144, 145, 147–149, 153–155, 186, 188
Syrjämäki, J., 172, 176

Takimoto, A., 39, 47, 207, 213, 215, 216, 226
Tanaka, Y., 148, 156
Tauber, C. A., 220, 228
Tauber, M. J., 220, 228
Taylor, D. H., 98, 115
Taylor, J. L., 158, 166
Tazawa, M., 18, 30, 74, 80, 82, 85, 86, 87
Teorell, T., 151, 156
Terracini, E. D., 164, 166
Thach, B., 69
Thibault, C., 228
Thomas, Ch., 141, 228
Thomas, R., 53, 69
Thomas, R. G., 219, 228
Tien-Hu Tsai, 194
Till, J. E., 192, 194
Todt, D., 37, 38, 47, 231, 236, 239
Tongiorgi, P., 100, 115
Tournois, J., 196, 228
Tribukait, B., 107, 115
Trucco, E., 148, 154
Truman, J. W., 54, 55, 66, 69

Tschudy, P. D., 143, *156*
Tuffli, C. F., *155*
Tukey, H. B., 157, *166*
Tuppy, H., 149, *153*
Tweedy, D. G., 10, *32*, 93, 99, *115*
Tyschenko, V. G., 54, *66*, *224*

Uebelmesser, E. R., 8, *32*
Umrath, K., 121, 127, *134*
Underwood, H., 62, *69*, *227*
Ungar, F., 50, 58, *68–70*, *194*
Ushatinskaya, R. S., 10, *32*
Utkin, I. A., 144, *156*

Vallbona, C., *165*
Van den Driessche, Th., 39, *47*, 145–148, *156*
van Pilsum, 135, *153*
Vasama, R., 145, *156*
Vasil'ev, V., *165*
Vedral, D. F., *194*
Venter, J., 135, *156*
Vielhaben, V., *133*, *153*, 185, 186, *188*
Visscher, M. B., *47*, *67*
Völker, H., 56, *70*
Volm, M., 43, *47*

Wagner, E., 152, *156*
Wagner, R., 83–85, *88*
Wahl, O., 7, 21, *32*, 72, *88*, 160, 161, *166*
Walcott, Ch., 169, *176*, *228*
Walkey, D. G. A., 8, *32*
Wall, J. R., *113*
Wallraff, H. G., *155*, 173, *175*
Warren, D. M., 9, 10, *32*, 140, *156*
Wassermann, L., 39, 41, 43, *47*
Waxman, A., *156*
Webb, H. M., 67, 80, 87, *165*, 187, *194*
Webb, O. L., 35, *46*, *194*
Weber, F., 107, 110, *114*, 144, *156*
Wegmann, H., *114*, *237*
Weight, F., *113*
Weismann, A., 198, *228*
Wellensiek, S. J., 219, *228*

Wellso, S. G., 219, *228*
Welsh, J. H., 16, 17, *29*, 51, 52, *70*
Went, F. W., 157, 165, *225*, 236, *238*
Werner, G., 138, *156*
Wever, R., 11, 12, 20, *28*, *29*, 56, *86*, 93, 94, *113*, *115*, 116–118, *133*, 163, *166*
Whitson, G. L., *29*
Wigglesworth, V. B., 52, *66*
Wilander, O., *66*
Wilkins, M. B., 9, 10, *29*, *32*, 40, *46*, 49, 70, 72, 83, *88*, 99, 109, *112*, *115*, 140, *156*
Williams, B. G., 52, 181, *188*, 204, 217, *228*
Wilson, W. O., 60, *66*, *69*, 89, *113*, *227*
Winchester, D. P., 222, *228*
Winfree, A. T., 39, *47*, 97, *115*, 131, *134*
Wintersberger, E., 149, *153*
Withrow, R. B., *46*, *65*, *225*
Wobus, U., 93, *115*, 132, *134*
Wodzicka-Tomaszewska, M., *225*, *226*
Wolf, J., 10, *32*
Wolf, W., *6*
Wolfson, A., 199, 215, 219, 222, *225*, *228*
Wollgiehn, R., 146, *153*
Woodward, D. O., *32*
Wurtman, R. J., 1, *6*, 61, 62, *66*, *70*

Yamamoto, Y., 52, *67*
Yanagishima, S., *31*
Yeates, G. W., 25, *32*
Yorke, 63
Young, J. Z., 56, *70*
Yugari, Y., 50, *67*

Zeno, J. R., *165*
Zeuthen, E., 13, 14, *29*
Ziegler Page, J. Z., 186, *188*
Zimmer, R., 104, 105, *115*, 128, *134*
Zimmerman, W., 98, *113*, *115*
Zinn, J. G., 15, *32*
Zucker, I., 61, *69*
Zusy, F. D., *32*
Zweig, M., *69*, 98, *115*

Subject Index

accuracy of periods, 18
—, time sense, 168
—, photoperiodic control, 201
Acetabularia, 39, 143, 145 ff
Acheta, 52
Acronycta, 23, 200
ACTH, 50, 57 ff
Actinia, 177
actinomycin, 145 ff
action current, 124 ff
action spectra, 41, 99, 215 ff
adrenal, 50, 57 ff, 75, 135
Agrolimax, 79
airplane flight, 3, 93, 100, 230 ff
alcohol, 140, 151
algae, see *Acetabularia, Chara, Chlamydomonas, Dictyota, Enteromorpha, Euglena, Gonyaulax, Halicystis, Hantzschia, Hydrodictyon, Oedogonium*
Allium, 143
all-or-none responses, 120 ff
Amaranthus, 94
amino acids, 136
—, see also histidine, leucine
amphibians, 98
—, see also frog
annual rhythms, 3, 27, 220 ff
Antheraea, 55, 204, 218
ants, 167, 173
aphids, 199 ff
Aplysia, 50, 142, 152, 183, 190
arctic regions, 24–26, 93, 109, 110, 163, 171 ff, 232
—, see also South Pole, Spitsbergen
Arctosa, 80, 171 ff
arrhythmicity, 39, 53

Ascobolus, 14
ascorbic acid, 58
asymmetry, 119 ff
atmospheric pressure, 163
ATP, 143
Avena, 18, 41, 80, 138 ff
Averrhoa, 121
awakening, 4, 17, 75, 161
Axolotl, 60
azimuth, 167 ff

bacteria, 149, 150
barbituric acid, 60
bats, diurnal processes, 19, 36, 75
—, reproductive cycle, 201
Bauhinia, 99
bean, see *Phaseolus, Vicia*
beat phenomena, 184, 185, 234
beech, 204
bees, time-sense, 2, 7, 16, 21, 72, 138, 159 ff
—, orientation, 167 ff
betacyanin, 152
birds, direction finding, 167 ff
—, light receptors, 98, 214
—, migration, 199 ff
—, reproduction, 60, 197 ff
—, synchronizers, 163
Blaberus, 132
Blennius, 177
blocks, physiological, 197 ff
blood, circadian rhythms, 57 ff
blood parasites, 62
blood pressure, 55
blue light, 41, 98, 189
body temperature, 17, 19, 56, 60, 75

251

bollworm, 183, 219, 221
Bombyx, 200
brain, rhythmic activities, 50, 52 ff
—, light absorption, 98, 217
—, tumor, 235
breeding periods, 199 ff
Bryophyllum, 10, 42, 49, 83, 99
buds, 204, 221
bullfinches, 89

Ca^{++} excretion, 11, 235
cabbage butterfly (Pieris), 202 ff, 218
Calandra, 183, 190
callus, 49
Canavalia, 8, 107, 108
cancer (carcinoma, tumor), 192, 235
Carabus, 52, 144, 209
Carausius, 53
carcinoma (cancer, tumor), 192, 235
Carcinus, 84, 178 ff
carotinoids, 99
cave animals, 26
cell division, 13, 40, 43, 48, 74
—, see also mitosis
cell synchronization, 13
Cervus, 199
Cestrum, 220
Chara, 129
chemical effects on photoperiodism, 197
— on rhythms, 138 ff
Chenopodium, 35, 92, 152
chicken, activity, 22, 43, 220
—, liver glycogen, 43
—, ovulation, 207
children, 22, 44
chilling experiments, 80 ff, 118 ff
Chlamydomonas, 24, 48
Chlethrionomys, 25
chloramphenicol, 148
chlorophyll, 41, 99, 146, 190, 204
chloroplasts, 39, 145 ff
Chorthippus, 98
chromatophores, 51
—, see also chloroplasts, pigments, plastids
chromosomes, 24
Chromulina, 177
chronon, 148
Chrysopa, 220
Cichorium, 36, 37, 38, 236

circadian, definition, 1
circadian organization, 63, 64
circadian rhythms, discovery, 14–17
circadian rule, 89, 90, 106
circannual rhythms, 27, 220 ff
Cl^- excretion, 11, 56
Clunio, 179, 184
cockroaches, see Blaberus, Leucophaea, Periplaneta
colchicine, 138
Coleus, 99, 105, 106
Colorado beetle, activity, 100
—, photoperiodism, 204
CO_2 metabolism, 10, 42, 49, 54, 56, 83, 147, 211
—, effects, 137
—, see also photosynthesis
color changes, 52, 199
—, see also pigment dispersion
compound eyes, 98
constancy of periods, 18–20
Convoluta, 177
corpora allata, 52 ff, 144
— cardiaca, 53, 54
corticosterone, 51 ff
Coryphoblennius, 179
cosmic radiation, 163
coupling of oscillators, 100, 124, 172
coupling to the clock, 11, 23, 44, 171 ff, 197
cricket, 52, 54
critical day length, 23, 199 ff
cues (= synchronizers, Zeitgeber), 12, 162 ff
—, unknown, 12, 163
cyanide, 136, 137
cycloheximide, 136

D_2O, 140, 151, 152, 181
damping, 34 ff, 120 ff
Daphnia, 189, 201
dark periods, initiating effects, 39 ff, 122
—, role in photoperiodism, 204 ff
Daucus, 50
day length, influence on development, 3, 196 ff
—, — on sun-orientation, 173
—, — on synchronization, 93
DCMU, 136
Desmodium, 127, 128

developmental cycles, 13, 14
diapause, 23, 199 ff
Dictyota, 183 ff
dinitrophenol, 136, 140
direction finding, 167 ff
diseases, 229 ff
disphasia, 99
dissociation, 63, 64, 99, 172, 211, 231 ff
DNA, 145 ff
DNP, 136, 140
domestication, 26, 197
dormancy, 3, 198 ff
Drosophila, 10, 21 ff, 34 ff, 51, 71 ff, 80, 89, 97 ff, 120, 138, 158, 230
drugs, susceptibility to, 191
ducks, 217
duckweed, 211
dung-beetle, 21, 92
duodenum, 193, 234

eodysis, 54, 55
egg laying, 98, 199, 207, 219
— release, 183
electric fields, 163
electric rhythms, 128, 129, 142, 190
electrographs, 190
electro-shock, 13
Enteromorpha, 183
entrainment, 79 ff
entrainments, limits, 80, 107
enzymes, 58, 135 ff
eosinophil rhythm, 38, 57 ff
Ephyppiger, 54, 98
epidermis, 204
equator, 197, 201
Escherichia, 192
Eskimos, 25
estrus-cycle, 62
ethyl alcohol, 140, 151, 152
Euglena, metabolic rhythms, 135 ff
—, phototaxis and mobility, 48, 73, 80, 108, 109, 136 ff, 189
—, tidal rhythm, 177
Eurydice, 179, 182
ewes, 220
Excirolana, 140, 179, 180, 184, 186
excitation substance, 125
excretion, see urinary excretion, Ca^{++}, K^+, Na^+
extraretinal light absorption, 98, 214

eyes, rhythmic processes, 142, 152, 190
—, role in phase shifts, 97 ff
—, role in photoperiodism, 214 ff
—, sensitivity to light, 190
eyestalks, 52

fade-out, 34 ff, 120 ff, 236
far-red, 36, 90, 105, 106
fasting, 50
ferret, 205
fiddler crab, see *Uca*
Filaria, 36
fire fly, 17
fishes, light receptors, 98
—, orientation, 169 ff
flagellates, see *Chlamydomonas, Chromulina, Euglena, Gonyaulax*
flatworms, 177
flavins, 99
flight, see airplane flight, space flight
flowering, 195 ff
flowers, nectar-secretion, 3, 159
—, odor-production, 159
—, opening and closing, 3, 7, 8, 11, 36, 49, 74
—, pollen offering, 159
flying squirrel, 19, 102, 103
Formica, 167, 173
fowl, 219
frequency, changes, 18, 19, 110
—, demultiplication, 110, 141
—, see also length of periods
frogs, 62, 169
frontal ganglion, 54
fungi, light effects, 99
—, sporulation, 8, 14
—, temperature compensation, 150
—, zonation, 13, 14, 41
—, see also *Neurospora, Pilobolus*
fur color, 199

gametes, release, 158, 183 ff
gastric acid, 234
— ulcer, 233, 234
gastrointestinal hormone, 234
gates, 14, 145
geotaxis, 172
Geotrupes, 21, 92

Subject Index

Glaucomys (= flying squirrel), 19, 102, 103
Glis, 36
Glycine, 51, 96, 210
—, see also soybean
glycogen, 13, 16, 43, 57, 192
Golgi bodies, 149
gonadotrophin, 61
gonads, 199 ff
Gonyaulax, 10, 38, 42, 48, 73, 74, 80, 99, 135, 136, 142, 144 ff
grasshopper, 54, 98, 219
ground squirrel, 75
growth, 15, 49, 80, 139, 140, 157
grunion fish, 183
Gryllus, 21, 54

Halicystis, 186
hamster, 9, 49, 50, 75
Hantzschia, 177
harmonic oscillations, 119 ff
head clock, 4, 17, 19, 161, 162
heat resistance, 190
heavy water, 140, 159 ff, 181
Heliothis, 183, 219, 221
heredity, 15, 20 ff
hibernation, 35, 36, 75, 76
—, see also diapause, dormancy
Hill reaction, 147
histidine, 50
horae variae resistentiae, 193
hormones, 51 ff, 138
hourglass principle, 2, 207 ff
humans, air flight, 3, 230 ff
—, behavior in the arctic, 25, 93, 163, 232
—, diseases, 229 ff
—, diurnal rhythms, 11, 17, 55 ff
—, drug-application, 193
—, head clock, 4, 17, 19, 161, 162
—, liver activity, 233
—, night-work, 3, 231 ff
—, responsiveness to external factors, 189 ff
—, sleep, 44
—, space travel, 4, 163, 164, 230
—, synchronizers, 163
—, with wrong clocks, 232
humidity, 158
Hyalophora, 55
hydrostatic pressure, 182

Hyoscyamus, chlorophyll formation, 147, 208
—, photoperiodism, 208
hypnosis, 4, 161, 162
hypophysis, 50, 56 ff
hypothalamus, 61, 62, 144

individual pattern, 19, 20
infants, 22, 44
initiation of rhythms, 34 ff
insulin, 193
intestine, 49, 50, 75
inversion experiment, 92, 163
Isospora, 62

Junco, 199, 206, 219

K+ excretion, 11, 25, 56, 64, 232, 233
K+ transport, 150 ff
Kalanchoe, flowering, 202, 208 ff
—, heat resistance, 190
—, movements, 8, 41, 74, 79, 104, 124, 144
katydid, 54, 98
KCN, 140
ketosteroids, 57 ff
kidneys, 38, 193, 235

Lacerta, diurnal rhythms, 19, 73, 90
—, direction finding, 172, 173
—, see also lizards
Lampetra, 56
landmark, 173
Lapland, 24, 171
Lasius, 167
latitude, and photoperiodism, 199, 200
—, and sun-orientation, 172
—, see also arctic regions, South Pole, tropical regions
leafcutter bee, 99
leaf movements, diurnal, 8 ff, 14 ff
—, see also Canavalia, Chenopodium, Coleus, Phaseolus
lemming, 50
Lemna, 211
Leiobunum, 50

Subject Index

length of periods, accuracy, 18, 19
—, changes, 18, 19, 110
—, chemical effects, 139 ff
—, heredity, 15, 21 ff
—, light effects, 89 ff
—, temperature effects, 71 ff
—, variability, 18, 19
Leptinotarsa (potato beetle), activity, 100
—, photoperiodism, 204
leucine, 144
Leucophaea, 79, 82, 84
Leuresthes, 183
life duration, 230
light, continuous, 34 ff, 89 ff, 157, 236, 237
—, effect on phase and period, 92 ff
—, initiating effect, 39 ff
—, perception, 97 ff
—, quality, see blue, far-red, red
—, see also day length
light breaks, effect on rhythms, 100 ff
—, photoperiodic effect, 205 ff
Ligia, 8, 9, 107, 112
lithium, 152
liver, 13, 16, 43, 135, 144, 152, 233
lizards, activity, 19 ff, 73, 80, 158
—, orientation, 169, 172, 173
—, photoperiodism, 220, 222
lobi optici, 53, 55
luciferase, 10, 135
luminescence, 10, 13, 17, 48, 73, 74, 80, 99, 136
lunar cycles, 177 ff
lunar orientation, 173, 174
luteinizing hormone, 60
lymphocytes, 64

Man, see humans
master clock, 8
mathematical approaches, 116 ff
mating behavior, 12, 48, 145, 158
Megachile, 99
Megoura, 201, 207 ff, 217, 218
melanophores, see pigment dispersion
melanophoric hormone, 52
membrane fluxes, 50, 151 ff
membrane oscillations, 151 ff
menstrual cycle, 62
Mesocricetus, 9, 50

Metriocnemus, 50, 201
Mice, activity rhythms, 19 ff, 75, 89
—, adrenal cycle, 58 ff
—, effect of poisons, 192
—, eosinophil rhythm, 38
—, mitotic cycles, 58, 145
Micrasterias, 189
Microfilaria, 62
migration, 197 ff
mitochondria, 149 ff
mitosis, 11 ff, 38, 49, 58 ff, 144 ff, 192
—, see also cell division
models, 116 ff
molting, 199
mRNA, 141
muscles, 124
mussels, 177, 183
mustard gas, 192
mutants, 23, 24
mutual entrainment, 51
Myotis, 19, 36, 75

N_2, 139
Na^+ excretion, 56, 64, 232
NaCN, 136
NaF, 136
narcosis, 138
nectar secretion, 3, 159
nephronic groups, 38
nerves, 51 ff, 98, 124, 128, 131, 183, 217
neuro-secretion, 50 ff, 144
Neurospora, 13, 14, 24
night-moth, 21
night-work, 3, 93, 231
Niphargus, 26
Nomadacris, 219
nucleid acids, 141, 145 ff
nucleolus, 52, 144, 145
nucleus, 143 ff
—, division, see mitosis
—, volume changes, 39, 51, 52, 143 ff, 149

O_2, consumption, 10, 26, 56, 75
—, withdrawal, 137 ff
oat, see *Avena*
ocelli, 98
odor production, 159

Oedogonium, bleaching, 237
—, sporulation, 18, 35, 43, 74, 79 ff, 137, 138
Oenothera, 236
optic lobes, 53, 55
optic nerve, 51
orangutan, 162
Orchestoidea, 174
Orconectes, 26
organ cultures, 49 ff
orientation, 167 ff
oviposition, 98, 199, 207, 219
ovulation, 60, 197 ff
oxygen, see O_2

Palaemon, 182
palolo-worm, 183
parabiosis, 53
Paramecium, 12, 39, 43, 99, 145, 146, 158
parieto-visceral ganglion, 50
pars intercerebralis, 52
Passer, 62, 209, 214
pathological phenomena, 229 ff
pea, 157
peanuts, 157
Pectinophora, 200, 201, 215
Pedaerus, 173
pendulum oscillation, 117 ff
penguin, 25
period, see length of periods
Periplaneta, 27, 53, 73, 82 ff
permeability, 152
Peromyscus, 75, 89, 90
Phaleria, 171
phase, definition, 12
phase-response curve, 77 ff, 100 ff, 123 ff
Phaseolus, 8 ff, 18 ff, 35, 39, 42, 71 ff, 101 ff, 119 ff, 151 ff, 214
pheromones, 158
Phormia, 230
Photinus, 17
photoperiodism, 3, 23, 196 ff
photosynthesis, 9, 38, 48, 135, 136, 145 ff
phototaxis, see *Euglena*
phytochrome, 99, 216 ff
Pieris, 202 ff, 218
pigeons, 169 ff

pigments, dispersion (migration), 8, 16, 35, 51 ff, 56, 80, 107, 112, 184
—, role in initiation, 42
—, role in phase shifts, 98, 99
—, role in photoperiodism, 216 ff
Pilobolus, 237
pineal gland, 61, 62, 98, 144
pituitary, 56 ff, 220, 235
plant lice, 199 ff
Plasmodium, 62
plastids, 147 ff
—, see also chloroplasts
Platynereis, 51, 183, 186
poisons, effect on rhythms, 136 ff
—, susceptibility to, 191 ff
pole, 163
pollen, 3, 159
pond skater, 82, 171, 172
potato beetle, activity, 100
—, photoperiodism, 204
preparatory phase, 219 ff
prey catching, 158
Procambarus, 98
prolactin, 60, 193
proteins, 135
pulse frequency, 17, 55, 56
puromycin, 136, 148
Pygoscelis, 25

Q_{10} values, 71 ff, 142
Quelea, 197
quinine, 138

radiosensitivity, 192
Rana, 62
rats, diurnal rhythms, 22, 59 ff, 80, 89 ff, 135, 143, 144, 152
—, estrus cycle, 60
—, periodic disease, 235
red, effect on rhythms, 36, 41, 90, 91, 105, 106, 163, 189
—, photoperiodic effect, 215 ff
refractory periods, 121 ff
—, in photoperiodism, 219 ff
relaxation oscillation, 117 ff
reproductive cycles, 197 ff
respiration, 50, 138 ff
—, see also CO_2 metabolism, O_2 consumption

Subject Index

response curves, see phase-response curve
rest periods, see diapause, dormancy
rice, 201
rifampicin, 148
RNA, 141, 145 ff
Romalea, 54

Saccharum, 221
salivary glands, 51
Salmo, 158, 191
sandhopper, 170, 174
Sceloporus, 158
sea anemone, 177
sea hare, 50, 142, 152, 183, 190
sea urchin, 183
seismonasty, 125
Sempervivum, 196
Serinus, 199
Sesarma, 179
sex glands, 199 ff
sexual merging, 158
—, see also mating behavior
sexual reactivity, 12, 48, 145, 158
sheep, 199, 220
shift work, 3, 93, 231 ff
shock-phase hypothesis, 236
shore crab, 178 ff
silkworm, 54, 55, 200, 205
sinus gland, 52
skeleton photoperiods, 213, 215
sleep, 17, 44, 45, 157
snail, 79
social synchronizers, 163
sound, 163
South Pole, 163
soybeans, 25, 51, 96, 157, 210, 213, 230
space flight, 4, 163, 164, 230
sparrow, blood parasites, 62
—, light absorption, 98, 214
—, locomotor activity, 62
—, reproductive cycle, 207 ff
Spitsbergen, 25, 232, 234
splitting, 63, 64, 142, 182
sporulation rhythm, 8, 14, 35, 43, 74, 79, 80, 137, 237
starch grains, 147
starling, 99, 167 ff, 217
starvation, 13, 135
steric alteration, 135

stridulation, 54, 98
subesophageal ganglia, 52 ff
subtle factors, 163
sugar cane, 221
sun-compass, 3, 167 ff
Synchelidium, 180
synchronization, mutual, 51
synchronized cells, 13
synchronizers, 12, 162 ff

Talitrus, 170, 174
Tamarindus, 14
Teleogryllus, 54, 98
temperature, changes in the body, 17, 19, 56, 60, 75
—, coefficients, 71 ff
—, effect on critical day length, 201 ff
—, effect on development, 157
—, effect on rhythms, 71 ff
—, initiating effect, 43
temperature resistance, 201 ff
Tenebrio, 132
theobromine, theophylline, 140
thermoperiodism, 157, 191, 229
thigmonasty, 125
thymidine, 193
thyroid, 59 ff, 235
thyroxine, 138
tidal cycles, 177 ff
time-sense, 2, 7, 16, 21, 71, 80, 137
tissue cultures, 49
tomatoes, 157, 230, 236
transients, 71 ff, 77 ff, 102, 117 ff, 123 ff
transocean experiment, 161
tropical plants, 76
tropical regions, 27, 172, 197, 221
tryptamine, 50
tumor (cancer), 192, 235
Tupaia, 64, 233
turkeys, 217, 219
turtle, 169
twilight, 109, 204

Uca, diurnal rhythms, 35, 80, 190
—, tidal rhythm, 177 ff
ulcer, 233, 234
ultrastructural changes, 149
ultraviolet, 99, 146

urethane, 138
urinary excretion, 17, 25, 56 ff, 64, 99, 231, 232

Valinomycin, 150 ff
*V*elia, 82, 171, 172
vernalisation, 218
*V*icia, 42
vitamins, 138

walking stick insect, 52, 53
water, excretion, see urinary excretion
—, pressure, 182
—, role in rhythmicity, 151, 152
—, turbulence, 182
weasel, 199

weaver finch, 197
wilting, 152
wolf spider (= *Arctosa*), 80, 170 ff
wood beetle, 26
wool-growth, 199, 220

X-chromosome, 24
X-rays, 193

yeast, 141

Zeitgeber (= cues, synchronizers), 12, 15, 163, 164
—, unknown, 12, 163
Zonotrichia, 62, 98, 207 ff, 219